Indicator Plants
of Coastal British Columbia

Indicator Plants
of Coastal British Columbia

K. Klinka, V.J. Krajina,
A. Ceska, and A.M. Scagel

Sponsored by

UBC PRESS / VANCOUVER

Printed in Canada on acid-free paper ∞

ISBN 0-7748-0321-5

Canadian Cataloguing in Publication Data

Main entry under title:
Indicator plants of coastal British Columbia

       Includes bibliographical references.
       ISBN 0-7748-0321-5

       1. Plant indicators – Pacific Coast (B.C.) 2. Botany – Pacific Coast
(B.C.) – Ecology. I. Klinka, K., 1937-
QK203.B7I53   1989   581.5'222'097113   C89-091062-6

Funding for the development of this guide was provided initially by the
Research Program of the B.C. Ministry of Forests. Project completion and
publication were funded, in part, by the Canada-British Columbia Forest
Resource Development Agreement – a five-year (1985-1990) $300 million pro-
gram cost-shared equally by the federal and provincial governments.

UBC Press gratefully acknowledges the ongoing support to its publishing program from the Canada Council, the Province of British Columbia Cultural
Services Branch, and the Department of Communications of the Government
of Canada.

Design and typesetting: T.D. Mock and Associates Inc., Victoria

UBC Press
University of British Columbia
6344 Memorial Road
Vancouver, BC V6T 1Z2
(604) 822-3259
Fax: 1-800-668-0821
E-mail: orders@ubcpress.ubc.ca

CONTENTS

FIGURES

ACKNOWLEDGEMENTS

We appreciate the continuing interest of many foresters in the development of an illustrated reference manual on indicator plants. We wish to thank R. Scagel, Pacific Phytometric Consultants, Vancouver, B.C., for co-ordinating the initial work; P. Courtin, B.C. Forest Service, Vancouver Forest Region, for data processing; G. Moyer, Victoria, B.C., L. Cuthbertson, G. Kayahara, A. McGee, and A. Pearson, Faculty of Forestry, University of British Columbia, for editing and preparing the manuscript for publication; and P. Quay, Faculty of Forestry, University of British Columbia, for typing the manuscript. We are also grateful to Dr. D. Minore, U.S. Dept. of Agric., Forest Service, Pacific Northwest Research Station, Dr. J. Kimmins, Faculty of Forestry, University of British Columbia, and Dr. J. Pojar, B.C. Forest Service, Prince Rupert Forest Region, for reviewing the manuscript.

We express our great appreciation to F. Boas, R. Dickens, R. Long, R. Norton, J. Pojar, J. Rosenberg, N. Ross, R. Scagel, N. Turner, R. Turner, E. Underhill, W. van Dieren, K. Wade, J. Woolett, and J. Worrall who made available colour photographs for this guide.

Financial support for the study was provided by the British Columbia Forest Service, Research Branch (1977-1978), and by the Canadian Forestry Service and the B.C. Ministry of Forests and Lands under the Canada-British Columbia Forest Resource Development Agreement, Extension, Demonstration, Research and Development Sub-program (1987-1988). This support is gratefully acknowledged.

1

INTRODUCTION

There is more to a plant than its name, form, and attractiveness. Each plant species is adapted to a range of environmental conditions and is restricted to sites within this range. Knowledge of the ecology of plant species makes it possible to infer a site's qualities from the vegetation present. Using plants as site indicators is integral to applied plant ecology.

This guide draws on many studies describing plant-environment relationships and summarizes our present level of understanding of the ecological response of forest plants. It is designed to be a concise and coherent tool for all who wish to learn more about indicator plants of coastal British Columbia. If the user is able to find answers to questions about indicative values of species and without technical training interpret site quality from vegetation, this work will be serving its purpose. It is hoped that the guide, intended primarily for field foresters, will also be used by naturalists and students of biology, botany, ecology, geography, and soil science.

Background

At present there is an abundance of botanical information in technical texts (e.g., Hitchcock et al. 1955-1969; Schuster 1966-1974; Hultén 1968; Hale 1969; Lawton 1971; Hitchcock and Cronquist 1973; Taylor and MacBryde 1977; Scoggan 1978-1979), field guides, and popular handbooks (e.g., Lyons 1952; Szczawinski 1959, 1962; Taylor 1966, 1971, 1973, 1974a, 1974b; Schofield 1969; Porsild 1974; Taylor and Douglas 1975; Brayshaw 1976; Garman 1973; Soper and Szczawinski 1976; Underhill and Chuang 1976; Haskin 1977; Hosie 1979; Coupé et al. 1982; Angove and Bancroft 1983; Vitt et al. 1988). In all these books, however, ecological information is scant, often receiving treatment such as, a *species* grows in "open, dry woods."

Foresters and forestry students interested in ecosystem-specific management have requested information on the use of plants as site indicators. In North America such use has not advanced very far despite the great number of ecological studies and surveys conducted. In British Columbia, Bell (1971) made an initial attempt to use plants and groups of plants as site indicators. Klinka (1977) used plants as aids for estimating forest soil moisture and nutrient regimes in southwestern British Columbia. More recently, indicator values of plants relating to soil moisture and nutrients were proposed by Comeau et al. (1982), Pojar et al. (1982), and Klinka et al. (1984).

In British Columbia, V.J. Krajina and his students pioneered studies of plant-environment relationships. These studies were continued by the Ecological Program Staff of the B.C. Forest Service. In 1978, G.C. Warrack, Director of Research Division, invited Dr. Krajina to describe the ecological characteristics of forest plants of British Columbia. Dr. Krajina characterized nearly 3000 vascular plants (Krajina et al. 1986). This information has been expanded here for selected species, for site diagnosis in coastal British Columbia.

Scope

This guide includes 419 indicator species of coastal British Columbia. Most of these are forest plants but some are found primarily in non-forested ecosystems and others are only temporary inhabitants of cut-over areas. Some species occur sporadically, or even rarely, and have a restricted distribution; others are plentiful and are found throughout the region.

British Columbia can be divided into coastal (maritime or Pacific) and interior (continental or Cordilleran) regions. The transition between coastal and interior regions is a wide area that has been recognized by both climatologists and ecologists. This transition area (located on the leeward side of the coastal mountains) is included in the coastal region as its climate is moderated by the Pacific Ocean. This guide covers the whole coastal region (hypermaritime, maritime, and submaritime belts) and the subcontinental belt of the interior region of British Columbia (Figure 1).

Format

This guide presents those species that the authors consider to be the most useful indicators in coastal British Columbia. Specific terms and connotative symbols have been employed to present information succinctly, avoiding as much as possible, the use of technical terminology. In view of many guides on plant identification and a continuing technical training by the Ecological Program Staff of the B.C. Forest Service, descriptive text, keys, and other materials essential for plant identification have been omitted. Instead, the emphasis has been placed on the rationale for using indicator plants, how indicator plants are recognized, and demonstration of how indicator plants can be used in site diagnosis. The results are summarized in tabular form and in the abridged text.

It is recommended that the reader proceed through the guide step-by-step, understanding each section before proceeding to the next. The purpose and organization of the guide are covered in the first chapter. The basic concepts and the rationale for using indicator plants in forestry are discussed in Chapter 2. Concepts and methods related to indicator plants are dealt with in Chapter 3. In Chapter 4, the site attributes selected for plant indication are introduced and species with the same or similar indicator values are presented.

Chapter 5 describes three methods of indicator plant analysis recommended for site diagnosis. One should select methods suitable for specific objectives and particular situations. Unfortunately, the most objective and precise methods require extensive and time-consuming sorting and calculations.

Chapter 6 summarizes salient distributional and ecological characteristics for individual species. Common names are recorded if available (Meidinger 1987), but because relatively few species have universally recognized common names, Latin (or technical) names are always used in the text. Sources of scientific nomenclature are

Hitchcock et al. (1955-1969), and Scoggan (1978-1979) for vascular plants, Ireland et al. (1987) for mosses, Stotler and Crandall-Stotler (1977) for liverworts and hornworts, and Hale and Culberson (1970), Egan (1987), and Noble et al. (1987) for lichens.

The synopsis in Appendix I presents the life-form and recognized indicator values of each species. The species checklist in Appendix II includes common names, Latin names, and synonyms of scientific names, all of which are arranged alphabetically. Each main entry is followed by a page number containing the illustration and commentary for the species.

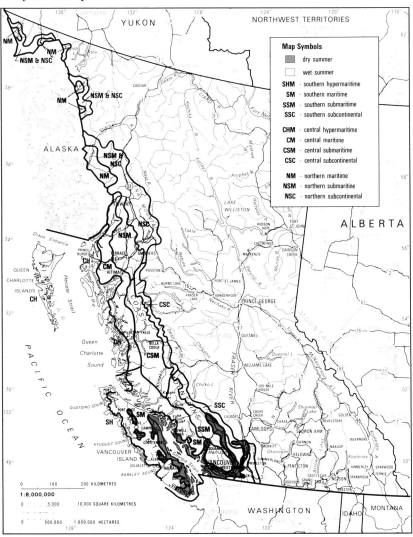

FIGURE 1   Approximate outline of precipitation (dry and wet) strata, longitudinal (hypermaritime, maritime, submaritime, and subcontinental) strata, and latitudinal (northern, central, and southern) strata used in classification of climates in coastal British Columbia

2

BASIC CONCEPTS

Ecosystem, Vegetation, and Site

In a forest community, there is a continual interaction among trees, understory species, and the environment. Attempts to understand plant adaptation and growth led to the development of the ecosystem concept. This concept implies that an organism cannot be considered separately from its environment, considers the ecosystem to be the basic functional unit of nature, and seeks to understand organism behaviour through the study of extremely varied and complex relationships between an organism and its environment.

In the geographical context, a forest ecosystem is a segment of landscape that is relatively uniform in climate, soil, plants, animals, and micro-organisms. Together, climate and soil, including topography, make up the site. Site is synonymous with environment and habitat. The biotic community of a site is composed of a combination of plants (vegetation), animals, and microorganisms, each of which forms its own community. Vegetation and soil are the two most obvious components of ecosystems.

Vegetation and soil integrate and reflect the influences of climate, topography, soil parent materials, organisms, and time on the ecosystem (Jenny 1941; Major 1951). Thus, on the basis of vegetation and soil, other ecosystem components can be assessed and individual ecosystems can be recognized and delineated.

The Ecosystem Concept in Forest Management

Sound and successful management of the forest resource results in forests that are more productive than natural forests. Such management requires knowledge of the forest's nature and characteristics. While rapidly developing technology has increased our ability to manipulate a stand, we have often ignored the ecological consequences of the manipulation (Van Dyne 1969). We are now beginning to appreciate, however, the importance of the ecosystem concept to forest management, especially in efforts to maintain or increase forest productivity.

Considering its relatively long production cycle, the managed forest must be fully stocked in order to obtain the specified benefits. All parts of the forest that are harvested should be replaced by new forest. To achieve full stocking a forester must decide on the most appropriate silvicultural treatments for many individual stands or sites. Decisions include tree species selection, site preparation, regeneration, stand tending, stand enhancement, and stand protection. Each decision emphasizes the need for detailed knowledge of site quality and vegetation-site relationships. Knowledge of vegetation is extremely valuable in assessing site quality and its temporal variation.

Many individual ecosystems occur within a forest. Despite the fact that no two ecosystems are exactly alike, not all are so different that they cannot be grouped on the basis of similarity in vegetation, sites, or both. Affinities among ecosystems exist and similar ecosystems usually occupy similar segments of the landscape. The application of the ecosystem concept in forestry is based on the premise that each ecosystem type requires a more or less specific set of treatments. Similar ecosystems, or those that are similar in vegetation and site, are expected to respond similarly to the same treatments.

Site Diagnosis

A forest site is a landscape segment that is uniform in climate and soil and has a certain vegetation potential. Both climate and soil are expressions of the integrated effect of many environmental factors, each of which directly or indirectly influences plants. Site quality is thus a particular combination of environmental factors expressed as a combination of climate and soil.

While it is fairly easy to enumerate site factors, it is difficult to evaluate their integrated effect on plants. The easiest way to tackle the problem is to consider vegetation. Vegetation plays the key role in the evaluation of site quality because:

1. it can be easily observed and objectively described; and

2. it is an integrator of the ecosystem the best expression of the combined influence of numerous environmental factors, the biotic community itself, and ecosystem history.

Indicator plants play a very useful but sometimes fallible role in site diagnosis. In some situations, (such as recently burnt sites,) vegetation may be absent. In other situations vegetation may be poorly developed, such as in young and dense forest stands where, due to very scanty light, only a few shade-tolerant mosses and liverworts are present. In most situations, however, understory vegetation is sufficiently developed so as to provide a reasonable indication of site quality despite differences in the floristic composition between early and late stages of succession on the same site.

Plants are always dependent on climate, soil moisture, and soil nutrients. Plants do not grow if the climate is too cold for transpiration, or it water is not available; and may grow poorly if essential nutrients are not in adequate supply. Therefore, climate, soil moisture, and soil nutrients have been used by plant ecologists to explain plant-environment relationships, and by foresters to characterize site quality. Sites with the same quality are expected to have similar vegetation and productivity potentials (Cajander 1926; Pogrebnyak 1930; Daubenmire 1968; Bakuzis 1969).

An ecosystem-specific approach requires recognition of individual ecosystems and identification of their basic site qualities: climate, soil moisture, and soil nutrients. For this purpose, Klinka et al. (1984) developed a field-based procedure termed site diagnosis, that Green et al. (1984) adapted for identification of site quality.

3

CONCEPTS AND METHODS RELATED TO INDICATOR PLANTS

Plants as Site Indicators

In plant communities species grow, reproduce, and survive or thrive under certain ecological conditions. The term niche is generally used to represent the complex of environmental factors (site quality) affecting each species. We can visualize a niche as a certain amplitude plotted in relation to one or more environmental factors (Rowe 1956; Bakuzis 1969; Krajina 1969; Shimwell 1972; Whittaker 1978). The lower limit of a factor, below which a species cannot grow, is called the minimum. The state associated with the species' best performance is called the optimum. The upper limit of a factor, beyond which the species is eliminated by excess, is called the maximum. The range between the minimum and maximum represents the ecological amplitude of the species (see Figures 2 and 3).

Once the ecological amplitude of a plant species is known, we can predict its performance on a particular site by determining the quality of the site. Conversely, if one knows the occurrence and vigour of a species at a particular site, it is possible to judge the quality of that site. This is the basic premise for using plants as site indicators. Each species has a more or less definite potential to indicate one or more site attributes that can represent single or complex environmental gradients or certain qualities. Whether or not a species is considered as an indicator depends on the criteria adopted for selecting the most useful indicator values.

Indicator plants provide site information faster than detailed physical and chemical measurements (Cajander 1926; Rowe 1956; Major 1969; Mueller-Dombois and Ellenberg 1974; Daubenmire 1976). While they are not intended to be substitutes for these measurements they do have many advantages. Because environmental factors may fluctuate over time, many measurements are required in order to evaluate their effects on plants. Furthermore, due to the interactions of physical and biological factors, we cannot interpret measurements pertaining to a single factor without considering the whole complex of factors. For example, to determine the moisture regime of a soil, we need to know not only the normalized precipitation and radiation, but also the information about demands for moisture and several soil properties that determine available water storage capacity. Plants, on the other hand, by responding to changes in ecological factors and integrating the effects of many individual factors, give a good and quick measure of the moisture regime encountered over a longer period than we can afford to monitor.

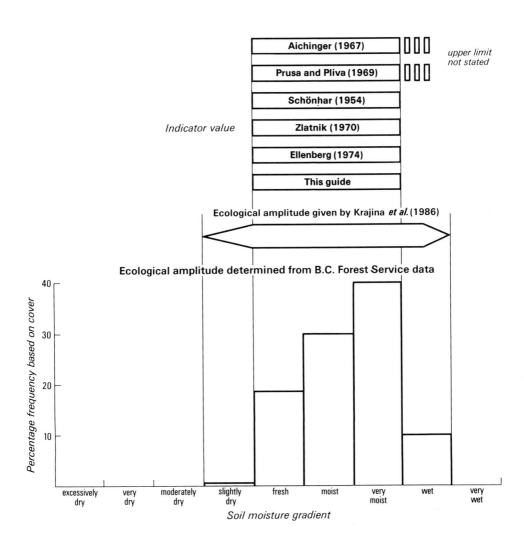

FIGURE 2   Ecological amplitude and indicator value of *Blechnum spicant* in relation to soil moisture as determined from B.C. Forest Service data and reported in several studies

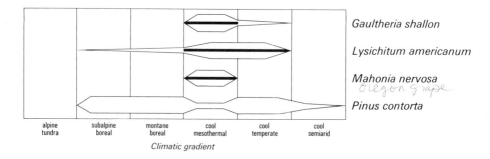

alpine tundra	subalpine boreal	montane boreal	cool mesothermal	cool temperate	cool semiarid

Climatic gradient

Gaultheria shallon

Lysichitum americanum

Mahonia nervosa
Oregon grape

Pinus contorta

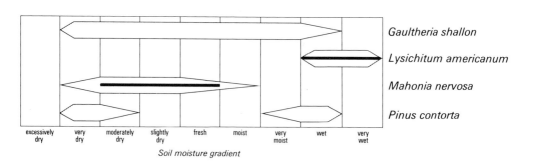

excessively dry	very dry	moderately dry	slightly dry	fresh	moist	very moist	wet	very wet

Soil moisture gradient

Gaultheria shallon

Lysichitum americanum

Mahonia nervosa

Pinus contorta

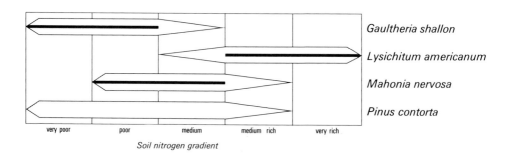

very poor	poor	medium	medium rich	very rich

Soil nitrogen gradient

Gaultheria shallon

Lysichitum americanum

Mahonia nervosa

Pinus contorta

FIGURE 3   Schematic ecological amplitudes (shown by diamond-like figures) and indicator values (shown by horizontal bars) of *Gaultheria shallon*, *Lysichitum americanum*, *Mahonia nervosa*, and *Pinus contorta* in relation to climate, soil moisture, and soil nitrogen

Choice of Objects of Plant Indication

The object of plant indication must be clearly stated and defined in order to arrive at consistent indicator values (Victorov et al. 1965). This object can be any biotic or environmental factor that influences the life of plants, such as a natural phenomenon (climate or soil) and its component properties (continentality of climates or soil moisture), or the kind and intensity of various conditions, processes, or events (a place where accumulated snow melts slowly, leaching of soil, or disturbance of an ecosystem by fire).

Environmental factors can be differentiated according to origin (climatic, topographic, and edaphic) or mode of action (light, heat, moisture, and nutrients) (Livingston and Shreve 1921). In the former group, one individual factor, within limits, can compensate for another factor. In the latter group, compensation does not take place. For example, a steep, south-facing slope may compensate for lack of heat in cool climates, or high precipitation may compensate for low water storage capacity in the soil. On the other hand, heat cannot replace light and moisture cannot replace nutrients in plant physiological processes. Consequently, similar vegetation may occur on morphologically different sites (sites different in terms of individual climatic, topographic, or soil factors), if these sites are similar in terms of light, heat, moisture, and nutrients. Thus plants are less reliable as indicators of individual site factors but more reliable as indicators of complex (synthetic) environmental factors.

We have chosen the following four site attributes - climate, soil moisture, soil nitrogen, and ground surface materials - as objects of plant indication. Each of these attributes is defined and classified in the section on Site Attributes. The classes or types distinguished for each attribute (e.g., alpine tundra is one of Köppen's climatic types used to characterize regional climates) provided a framework for describing ecological amplitudes and for calibrating indicator values of species.

Indicator Values of Species

The indicator value of a species is a statement on correlation between the species and its environment. The occurrence of such correlations merely suggests that different species are associated with different environments, but the correlations themselves do not suggest any clear cause and effect relationship (Whittaker 1954; Curtis 1959; Persson 1981).

Indicator values can be integers or names both referring to classes of qualitatively or quantitatively defined environmental gradients or certain site qualities. For example, Ellenberg (1974) has listed indicator values of approximately 2000 vascular plants in the western part of central Europe in relation to light, temperature, continentality, soil moisture, soil reaction, and soil nitrogen gradients, using, for each gradient, the same number of nominal classes – 1 to 9. For gradients used in this book, we have expressed indicator values by the names given to the classes of each gradient, but the number of our classes depends on the nature of the gradient. For example, soil moisture indicator values were derived from six soil moisture regimes recognized in coastal British Columbia by Klinka et al. (1984).

The indicator value of a species depends first on its ecological amplitude within its entire range, and second on its ecological and physiological response.

The only objective and quantitative way to assess the ecological amplitude of each species for each factor would be to measure a broad array of environmental factors in a large number of ecosystems. A large number of samples would have to be gathered in order to obtain representative data for all species in question. Such a task would re-quire a tremendous sampling effort; moreover, it would be almost impossible to create a proper sampling design and to interpret the results.

Analysis of vegetation data using the Braun-Blanquet tabular method is a reasonable means of obtaining ecological amplitudes (Mueller-Dombois and Ellenberg 1974; Westhoff and van der Maarel 1978). By organizing samples of vegetation into floristically similar groups at various levels of generalization, and characterizing these groups by common species, one arrives at so-called diagnostic species groups. By relating the diagnostic species groups to a set of measured environmental gradients, it is possible, through a process of successive approximation, to determine the ecological amplitudes of plant species. This approach is inherent in the ecosystem studies carried out from 1950-1975 by Krajina and his students in British Columbia and from 1975 to present by the B.C. Forest Service.

Krajina et al. (1986) provided the basic source of ecological amplitudes for the selected plant species. Much of this information was derived from the diagnostic species groups-environment relationships that are reported in ecological studies conducted by Krajina and his students in British Columbia (Szczawinski 1953; Arlidge 1955; McMinn 1957; Mueller-Dombois 1959; Archer 1963; Orloci 1961, 1964; Kuramoto 1965; Wade 1965; Krajina 1969; Brayshaw 1970; Brooke et al. 1970; Cordes 1972; Revel 1972; Wali and Krajina 1973; Beil 1974; Kojima and Krajina 1975; Klinka 1976). Whenever possible, the ecological amplitudes were checked against the data obtained by the Ecological Program Staff of the B.C. Forest Service and the indicator values given in North American and European studies (Taylor 1932; Schönhar 1952, 1954; Dahl 1956; Rowe 1956; Mezera 1957; Aichinger 1967; Minore 1969, 1972, 1979; Prusa and Pliva 1969; Zlatnik 1970; Jeglum 1971; Ellenberg 1974, 1982; Long and Turner 1975; Landolt 1977; Arbeitskreis Standortskartierung 1978; Bakuzis and Kurmis 1978; Tsiganov 1983). The procedure used to determine ecological amplitudes and indicator values is demonstrated in Figure 2. Due to different methods used in the characterization of attribute classes, it was often difficult to correlate precisely ecological amplitudes proposed by Krajina et al. (1986) and other workers. However, it was possible to develop general correlations, occasionally supplemented by detailed en-vironmental description of species or vegetation.

The ecological amplitude of a plant species depends on its competitive interaction with other plants in the community. For example, Ellenberg (1953) measured the growth of several plant species in pure cultures and found that in non-competitive conditions their 'physiological optima' were close to each other, and that all species grew best in fresh and more or less neutral (pH = 7) soil. When planted in mixtures with other species in the field, they grew their best under different soil conditions, expressing different 'ecological optima.' This shift from physiological optima to ecological optima was attributed to competition.

This shift has important implications in the determination of indicator values of species. Since different plants probably impose different competitive interactions species indicator values are usually described as being valid within certain geographic areas. A different flora may produce different competitive relationships, and hence produce different shifts in the ecological optima of species and produce different indicator species.

In some situations, competition may result in a bi-modal pattern in the ecological amplitude of a species. Some plants, such as lodgepole pine (*Pinus contorta*), occur in the coastal region either on very dry shallow soils on rock outcrops or in wet peat bogs. This bi-modal distribution results from competitive pressure on dry, fresh, and moist soils, where lodgepole pine can grow but is commonly outcompeted (Figure 3). Thus, the true ecological amplitude of lodgepole pine in relation to soil moisture includes the entire soil moisture gradient and its possible indicator value should be interpreted as very broad, ranging from very dry to wet soils.

Most plant species are normally distributed, occurring most frequently over a relatively narrow range and least frequently over relatively wide segments of their ecological amplitude. Sporadic occurrence and the low vigour of plants can be encountered at or near ecological minima and maxima of the species. That part of the ecological amplitude of a species that included approximately 85% of the observed occurrences was used to determine an indicator value of the species. Thus, in this guide, indicator values represent arbitrarily curtailed ecological amplitudes (Figure 3). At the extremes, presence of a species is sporadic, its cover and vigour are low, and its occurrence is usually limited to special microsites. For example, the species intolerant of water deficiency may grow on water-deficient sites on 'wetter' microsites (commonly in water-receiving depressions where the soil may be deeper and richer in organic matter than on mounds) and the species intolerant of a high water table may grow on waterlogged sites on 'drier' microsites (usually on mounds, stumps, or woody debris). However, this curtailment does not mean ignoring marginally occurring species in site diagnosis. Instead, it requires recording the presence and cover of all plant species on a site, if indicator plant analysis is to produce meaningful results.

Indicator values of species may be global or local. For example, *Anaphalis margaritacea* is considered a global indicator of exposed mineral soils accompanied by full light conditions; *Gaultheria shallon*, due to its sporadic occurrence in the interior wet belt (wet cool temperate climate), is considered a local indicator of cool mesothermal climates accompanied by acid organic substrates. Consequently, the indicator values for some species given in this guide may not be valid outside coastal British Columbia.

Choice of Indicator Species

Cluster analysis was employed to organize the data on indicator values of species. For each site attribute, the analysis grouped together species with the same or similar ranges of indicator values and produced several species groups. These groups were examined with respect to the range and overlap of indicator values among species groups. Obviously, species with consistently wide ranges are less definitive site indicators than those with consistently narrow ranges. Unfortunately, species with very narrow ranges are rare.

To make rational decisions on which species were the most suitable site indicators, the widest range of an indicator value was limited to three contiguous attribute classes. This criterion minimized the overlap of indicator values and allowed the use of a greater number of species as site indicators. Consequently, plant species with intermediate ranges were also accepted as site indicators; they occur commonly throughout the region and are still useful in site diagnosis. For example, *Gaultheria shallon* has a narrow range regarding climate, ground surface materials, and soil nitrogen (a good indicator of maritime cool mesothermal climates, Mor humus forms, and nitrogen-poor soils) and a wide range regarding soil moisture, extending from very dry to wet soils (a poor indicator of soil moisture) (Figure 3).

Indicator Species Groups

Regardless of life-form and geographic distribution, some species have the same or very similar indicator values. These species can be combined into an indicator species group (ISG) (Ellenberg 1950; Mueller-Dombois and Ellenberg 1974; Westhoff and van der Maarel 1978). Based on the cluster analysis described earlier, two or more ISGs were segregated for each site attribute (Table 1) and were characterized by the range of the indicator values of the included species.

Each ISG includes species with very similar and often identical indicator values but individual members of any group may differ in life-form, geographic distribution, and other indicator values. Therefore, the full complement of species of any group is not expected to occur together on any one site. Indicator species groups are presented in tables that include an alphanumeric symbol, the Latin name of the common and easily identifiable plant species used to represent the group, and an alphabetical list of species belonging to the group. The ISGs in this guide are tentative and subject to revision as additional information becomes available.

TABLE 1 Synopsis of recognized indicator species groups

Site attribute	Indicator species group (ISG)	
	Number of ISGs	Number of indicator species
Climate (CLIM)	6	235
Soil moisture (MOIST)	6	346
Soil nitrogen (NITR)	3	378
Ground surface materials (GSM)	5	289

4

SITE ATTRIBUTES AND INDICATOR SPECIES

Climate

Climate can be stratified by considering its regional, local, and plant community features (Major 1951). The plant community climate or microclimate is the climate within a plant community. Because it is a function of the vegetation, plant community climate can be characterized by the vegetation on a particular site. Local climates are affected by local topography but not by vegetative cover. Regional climates are affected neither by vegetative cover nor local topography. In order to characterize plant-climate relationships, we used regional climates and distributional patterns of species.

Köppen's climatic types, as amended by Trewartha (1968), were used to characterize different regional climates (Table 2). The classification of continentality types followed Meusel et al. (1965), Zlatnik (1970), Ellenberg (1974), and Tsiganov (1983) (Table 3). Boreal climates were divided on the basis of elevation and the length of growing season into high-altitude or subalpine (SB) and montane (MB) boreal climates.

Normal climatic data from a network of stations not topographically biased (Major 1963) and B.C. Forest Service biogeoclimatic maps were used to identify the climatic types, stratify vegetation data, and calibrate climatic indicator species. Continentality types were defined by using an index of continentality (Wilson and Rouse 1972):

$$1C = \left(\frac{1.7 \ [\text{mean T (July) - mean T (Jan.)}]}{[\text{sine (degrees latitude)}]} \right) - 20.4$$

where IC = index of continentality and T = temperature (°C).

TABLE 2   Key to the principal climatic types of British Columbia

Number	Characteristic	Climate type
1a	Excess of precipitation over evaporation	2
1b	Excess of evaporation over precipitation[1], mean annual temperature <18°C	**cool semiarid** (CSA)
2a	Mean temperature of the warmest month ≤10°C	**alpine tundra** (AT)
2b	Mean temperature of the warmest month >10°C	3
3a	Mean temperature of the coldest month ≤0°C	4
3b	Mean temperature of the coldest month ≷0°C	5
4a	Number of months with temperature >10°C is <4 (mean temperature of the warmest month <22°C)	**boreal** (B)[2]
4b	Number of months with temperature >10°C is ≥4 (mean temperature of the warmest month <22°C)	**cool temperate** (CT)[2]
5a	Mean temperature of the warmest month ≤22°C and number of months with temperature >10°C is <4	**cold mesothermal** (KM)
5b	Mean temperature of the warmest month <22°C, and number of months with temperature >10°C is ≥4	**cool mesothermal** (CM)[2] *Soleil*

[1] Evaporation exceeds precipitation when the mean annual precipitation (mm) ≤ ([0.44 (t°C + 17.78) 1/8] - 14) 25.4, where t = mean annual temperature [after Köppen in Trewartha (1968)].
[2] Mean precipitation of the driest summer month <30 mm **summer-dry** (SD)
Mean precipitation of the driest summer month ≥30 mm **summer-wet** (SW)

TABLE 3   Classification of continentality of climates of British Columbia

Index of continentality[1]	Continentality type and description *More Eastward*
<5	**hypermaritime** (HM) climates of the outer coast under the dominant influence of the Pacific Ocean
5 to <18	**maritime** (M) *Puget Sound* climates of coastal areas on the windward side of the coastal mountain ranges under the prevailing influence of the Pacific Ocean
18 to <26	**submaritime** (SM) *Port Townsend* climates of the coast-interior ecotone on the leeward side of the coastal mountain ranges under the strong influence of the Pacific Ocean
26 to <34	**subcontinental** (SC) climates of the coast-interior ecotone adjacent to the leeward side of the coastal mountain ranges under the weak influence of the Pacific Ocean
>34	**continental** (C) winter-cold and summer-warm to -hot climates of the interior areas east of the coastal mountain ranges under the prevailing influence of continental air masses

[1] Source of data: Klinka et al. (1979) and Courtin (B.C. Forest Service, Vancouver Forest Region, unpublished data).

Indicators of Climate

Precise information on climate is available for only a few sites - climatic stations. To obtain additional data would require long-term measurements, clearly a very expensive proposition considering the vast expanse of forest land. The indication of climate by plants, therefore, is a viable and valuable substitute for the lack of climatic data.

There are 235 indicator species in six ISGs for indication of climate in coastal British Columbia (Tables 5 through 10). Based on the indicator values adopted for diagnosing climate, each group represents a wide and over-lapping segment of the climatic gradient (Table 4). The ISGs are arranged according to accumulated growing degree days (over 5°C), which increase from the *Phyllodoce empetriformis*-group to the *Letharia vulpina*-group. Because of regional significance, we marked in Table 8 (the *Kindbergia oregana*-group) the species that occur prevailingly in summer-dry or summer-wet cool mesothermal climates. The species that commonly inhabits high-elevation sites (the *Phyllodoce empetriformis* ISG) occasionally descend to lower elevations where they can be useful indicators of late snow-melt sites.

Climatic indicators can be used to make inferences on regional climates, local climates, and microclimates. Indication of regional climate from the understory vegetation on a site, however, may be difficult due to the modifying effect of local topography and tree canopy on many climatic factors. For example, due to topographic influences the local climate of a site may deviate from the regional climate, i.e., it may be warmer or cooler, drier or wetter (from rain and/or snow), or milder or harsher. Biogeoclimatic maps, assembled by the B.C. Forest Service research program staff, facilitate identification of regional climates.

TABLE 4 Indicators of climate - synopsis of indicator species groups

Symbol	Name of ISG	Combination of climatic types
CLIM1	*Phyllodoce empetriformis*-group	alpine tundra and boreal (ATB)
CLIM2	*Chamaecyparis nootkatensis*-group	subalpine boreal and cool mesothermal (SBCM)
CLIM3	*Vaccinium membranaceum*-group	boreal and cool temperate (BCT)
CLIM4	*Kindbergia oregana*-group	cool mesothermal (CM)
CLIM5	*Holodiscus discolor*-group	cool temperate and cool mesothermal (CTCM)
CLIM6	*Letharia vulpina*-group	cool temperate and cool semiarid (CTCSA)

The assessment of continentality is particularly important in coastal British Columbia where north-south oriented insular and mainland mountain ranges impose windward and leeward climatic effects. For example, the sporadic occurrence of *Abies lasiocarpa* on some subalpine sites correlates with a higher continentality of the subalpine boreal climate. Similarly, the frequent occurrence of *Paxistima myrsinites* on many submontane to montane sites correlates with the less-maritime, cool mesothermal climate of eastern central and southern Vancouver Island. Thus, these and other plants which have a strong affinity to continental climates (the *Vaccinium membranaceum* and *Letharia vulpina* ISGs) can be used to indicate the degree of continentality of climate. In the coast-interior ecotone, occurrences of indicators of maritime climates decrease and those of continental climates increase along a longitudinal gradient.

TABLE 5   Indicators of alpine tundra and boreal climates

CLIM1	*Phyllodoce empetriformis*-group
Arnica latifolia	*Lycopodium alpinum*
Caltha leptosepala	*Lycopodium sitchense*
Cassiope mertensiana - M	*Petasites frigidus*
Cassiope stelleriana - M	*Phyllodoce empetriformis*
Cassiope tetragona - C	*Phyllodoce glanduliflora*
Cladina stellaris	*Pinus albicaulis* - C
Dicranum pallidisetum - C	*Polystichum lonchitis*
Epilobium latifolium	*Ranunculus eschscholtzii*
Erigeron peregrinus	*Rhododendron albiflorum* - C
Gaultheria humifusa	*Saxifraga tolmiei* - M
Hippuris montana	*Sibbaldia procumbens*
Leptarrhena pyrolifolia	*Stenanthium occidentale*
Luetkea pectinata	*Vaccinium scoparium* - C
Lupinus arcticus - C	*Vahlodea atropurpurea*

C - indicators of subcontinental to continental alpine tundra and boreal climates.
M - indicators of hypermaritime to submaritime alpine tundra and boreal climates.

TABLE 6   Indicators of hypermaritime to submaritime, subalpine boreal and cool to cold mesothermal climates.

CLIM2	*Chamaecyparis nootkatensis*-group
Abies amabilis	*Pilophoron clavatus*
Blechnum spicant	*Rhizomnium nudum*
Caltha biflora	*Rhytidiadelphus loreus*
Carex anthoxanthea	*Sorbus sitchensis*
Chamaecyparis nootkatensis	*Trichophorum cespitosum*
Cladothamnus pyroliflorus	*Tsuga mertensiana*
Coptis aspleniifolia	*Vaccinium alaskaense*
Cornus unalaschkensis	*Vaccinium deliciosum*
Fauria crista-galli	

TABLE 7   Indicators of boreal and cool temperate climates

CLIM3	***Vaccinium membranaceum*-group**
Abies lasiocarpa - C	*Pedicularis racemosa* - C
Antennaria neglecta - C	*Picea engelmannii* - C
Aralia nudicaulis - C	*Platanthera orbiculata* - C
Arnica cordifolia - C	*Pleurozium schreberi* - C
Aster ciliolatus - C	*Populus tremuloides* - C
Aster conspicuus - C	*Ptilium crista-castrensis* - C
Barbilophozia floerkei	*Pyrola chlorantha* - C
Barbilophozia lycopodioides	*Rhytidiopsis robusta*
Calamagrostis rubescens - C	*Rosa acicularis* - C
Clintonia uniflora	*Rubus idaeus* - C
Cornus canadensis	*Rubus pedatus*
Equisetum sylvaticum - C	*Rubus pubescens* - C
Gaultheria ovatifolia - C	*Salix bebbiana* - C
Geocaulon lividum - C	*Shepherdia canadensis*
Lathyrus ochroleucus - C	*Sorbus scopulina* - C
Lonicera utahensis - C	*Spiraea betulifolia* - C
Lycopodium annotinum	*Spiraea densiflora* - C
Lycopodium complanatum - C	*Stellaria calycantha* - C
Lycopodium obscurum - C	*Streptopus streptopoides*
Mitella breweri - C	*Tiarella unifoliata*
Mitella nuda - C	*Vaccinium caespitosum*
Mnium spinulosum - C	*Vaccinium membranaceum*
Parnassia fimbriata	*Vaccinium myrtilloides* - C
Paxistima myrsinites - C	*Viburnum edule* - C
Pedicularis bracteosa - C	

C - indicators of subcontinental to continental boreal and cool temperate climates.

TABLE 8   Indicators of cool mesothermal climates

CLIM4	*Kindbergia oregana*-group

Acer circinatum	*Lobaria oregana*
Acer macrophyllum	*Lonicera hispidula* - SD
Achlys triphylla	*Madia madioides* - SD
Agrostis aequivalvis	*Mahonia nervosa*
Aira caryophyllea - SD	*Maianthemum dilatatum*
Aira praecox - SD	*Malus fusca*
Alectoria vancouverensis - SD	*Mitella ovalis*
Alnus rubra	*Mycelis muralis*
Arbutus menziesii - SD	*Oemleria cerasiformis* - SD
Arctostaphylos columbiana - SD	*Oenanthe sarmentosa*
Bazzania tricrenata	*Perideridia gairdneri*
Boschniakia hookeri - SD	*Picea sitchensis* - SW
Boykinia elata	*Plagiomnium insigne*
Calamagrostis nutkaensis - SW	*Plagiothecium undulatum*
Camassia leichtlinii - SD	*Pogonatum contortum*
Camassia quamash - SD	*Polypodium glycyrrhiza*
Campanula scouleri - SD	*Polypodium scouleri* - SW
Cardamine nuttallii	*Polystichum munitum*
Carex hendersonii	*Prenanthes alata* - SW
Carex inops - SD	*Quercus garryana* - SD
Carex obnupta	*Rhizomnium glabrescens*
Carex sitchensis	*Ribes bracteosum*
Chimaphila menziesii	*Ribes divaricatum*
Circaea pacifica	*Ribes lobbii* - SD
Cornus nuttallii	*Ribes sanguineum*
Cytisus scoparius	*Rubus laciniatus*
Dicentra formosa	*Rubus spectabilis*
Disporum smithii	*Rubus ursinus*
Dodecatheon hendersonii	*Salix hookeriana*
Equisetum telmateia	*Sanicula crassicaulis* - SD
Erythronium oregonum - SD	*Satureja douglasii* - SD
Erythronium revolutum - SW	*Sedum spathulifolium* - SD
Festuca subulata	*Sisyrinchium douglasii* - SD
Festuca subuliflora	*Sphagnum papillosum*
Fritillaria lanceolata	*Sphagnum tenellum*
Gaultheria shallon	*Spiraea douglasii*
Gentiana douglasiana - SW	*Stachys cooleyae*
Gentiana sceptrum	*Stachys mexicana*
Geranium molle	*Symphoricarpos hesperius* - SD
Herbertus aduncus	*Tiarella laciniata*
Heuchera micrantha	*Tolmiea menziesii*
Homalothecium megaptilum	*Trillium ovatum*
Hookeria acutifolia - SW	*Usnea longissima* - SW
Hookeria lucens - SW	*Vaccinium ovatum* - SW
Isothecium stoloniferum	*Vaccinium parvifolium*
Kindbergia oregana	*Viola sempervirens*
Leucolepis menziesii	

SD - indicators of summer-dry cool mesothermal climates.
SW - indicators of summer-wet cool mesothermal climates.

TABLE 9 Indicators of cool temperate and cool mesothermal climates

CLIM5 *Holodiscus discolor*-group	
Abies grandis	Luzula multiflora
Adenocaulon bicolor	Melica subulata
Adiantum pedatum	Moehringia macrophylla
Allium acuminatum	Montia parvifolia
Allotropa virgata	Physocarpus capitatus
Angelica genuflexa	Pogonatum alpinum
Asarum caudatum	Pteridium aquilinum
Atrichum selwynii	Pyrola picta
Atrichum undulatum	Rhamnus purshianus
Cardamine breweri	Scapania bolanderi
Claopodium crispifolium	Scirpus microcarpus
Claytonia sibirica	Smilacina racemosa
Corylus cornuta	Taxus brevifolia
Elymus glaucus	Tellima grandiflora
Eriophyllum lanatum	Thuja plicata
Galium aparine	Trautvetteria caroliniensis
Hemitomes congestum	Trientalis latifolia
Holodiscus discolor	Tsuga heterophylla
Listera convallarioides	Viburnum trilobum
Lonicera ciliosa	

TABLE 10 Indicators of subcontinental to continental cool temperate and cool semiarid climates

CLIM6 *Letharia vulpina*-group	
Agropyron spicatum	Mahonia aquifolium
Ceanothus velutinus	Philadelphus lewisii
Disporum trachycarpum	Pinus ponderosa
Juniperus scopulorum	Prunus virginiana
Letharia vulpina	

Soil Moisture

Soil moisture regime (SMR) represents the long-term balance between the amount of available soil water and the demand for that water by vascular plants. Krajina (1969) adopted nine classes (0 to 8) of SMRs (so-called relative SMRs) and applied them consistently in each climate. For example, the relatively driest soil in any climate was always very xeric (0) and the wettest was always hydric (8). Despite ease in field identification (see Green et al. 1984), relative SMRs express the actual available soil moisture inconsistently so that a very xeric SMR in one climate could be drier or wetter than in another climate. It is inadequate to state that a soil is the driest in relation to other soils without referring to the magnitude of the water deficit.

To facilitate the use of plants as indicators of soil moisture, we used actual SMRs to characterize plant-water relationships. Klinka et al. (1984) proposed a tentative classification of actual SMRs for coastal British Columbia (Table 11) based on annual water balance (Figure 4) and the depth of the growing-season groundwater table (cf. Major 1963, 1977; Soil Survey Staff 1975). Actual SMRs for water-deficient sites were identified by extrapolating the results of soil moisture studies (Griffith 1960; McMinn 1961; Brooke et al. 1970; Roemer 1972; Klinka, Feller, and Lowe 1981; Spittlehouse and Black 1981; Giles 1983; Giles et al. 1985; Courtin, B.C. For. Serv., Vancouver For. Reg., unpubl. data), and for the sites with water surplus from data on the type, depth, and duration of the growing-season groundwater table. This information was used to stratify vegetation data and calibrate indicator species of soil moisture.

TABLE 11 Key to tentative actual soil moisture regimes for coastal British Columbia

Number	Soil moisture characteristic	Soil moisture regime
1a	Rooting-zone groundwater absent during the growing season	2
1b	Rooting-zone groundwater present during the growing season	5
2a	Water deficit occurs (soil-stored reserve water is used up and drought begins if current precipitation is insufficient for plant needs)	3
2b	No water deficit occurs	4
3a	Deficit >5 months or AET/PET[1] ≤40%	**excessively dry** (ED)
3b	Deficit >3.5 but ≤5 months or AET/PET ≤60 but >40%	**very dry** (VD)
3c	Deficit >1.5 but ≤3.5 months or AET/PET ≤90 but >60%	**moderately dry** (MD)
3d	Deficit >0 but ≤1.5 months or AET/PET >90%	**slightly dry** (SD)
4a	Utilization (and recharge) occurs (current need for water exceeds supply and soil-stored water is used)	**fresh** (F)
4b	No utilization (current need for water does not exceed supply)	5
5a	Groundwater table >60 cm deep	**moist** (M)
5b	Groundwater table >30 cm but ≤60 cm deep	**very moist** (VM)
5c	Groundwater table >0 but ≤30 cm deep	**wet** (W)
5d	Groundwater table at or above the ground surface	**very wet** (VW)

[1] AET - actual evapotranspiration, PET - potential evapotranspiration.

Indicators of Soil Moisture

Precise and site-specific information on soil moisture regime would require many measurements gathered over a period of three to five years, however, this is not a feasible proposition considering the vast amount of forest land and expense involved. Clearly, little indication is needed for wet soils where the water table is at or near the ground surface for most of the growing season, however, in other situations vegetation provides a useful indication of possible water deficit or surplus, or a fluctuating groundwater table.

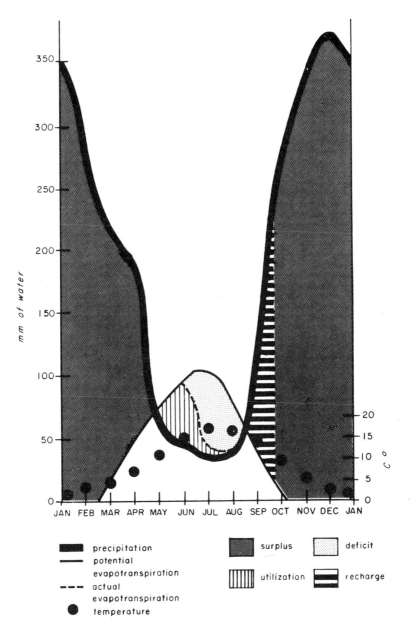

FIGURE 4   Annual water balance for moderately dry soil within summer-dry cool mesothermal climate. Available water-holding capacity is 90 mm.

There are 346 indicator species in six ISGs for indication of soil moisture in coastal British Columbia (Tables 13 through 18). Based on the indicator values adopted for diagnosing soil moisture (Table 12), each group represents a relatively wide and overlapping segment of the soil moisture gradient. In Tables 16 and 17, we marked the species that tolerate temporary flooding and thus can be useful indicators of sites with a markedly fluctuating groundwater table (poorly drained, fine-textured soils located on flats or very gentle slopes) or flooded sites (active alluvial floodplains). In areas where a high-winter and low-summer rainfall occur, actual SMR may fluctuate from winter-wet to summer-dry, significantly affecting seedling establishment and growth. The *Sphagnum*-group (Table 18) includes species that commonly inhabit waterlogged sites or wetlands (peat bogs, muskegs, fen, marshes, or swamps).

TABLE 12   Indicators of soil moisture: synopsis of indicator species groups

Symbol	Name of ISG	Range of actual SMRs
MOIST1	Lichen-group	excessively dry to very dry (EVD)
MOIST2	*Arctostaphylos uva-ursi*-group	very dry to moderately dry (VMD)
MOIST3	*Mahonia nervosa*-group	moderately dry to fresh (MDF)
MOIST4	*Blechnum spicant*-group	fresh to very moist (FVM)
MOIST5	*Rubus spectabilis*-group	very moist to wet (MW)
MOIST6	*Lysichitum americanum*-group	wet to very wet (WVW)

TABLE 13   Indicators of excessively dry and very dry soils

MOIST1   Lichen-group

Agropyron spicatum	*Eriophyllum lanatum*
Aira caryophyllea	*Juniperus scopulorum*
Aira praecox	*Luzula multiflora*
Cladina arbuscula	*Polytrichum piliferum*
Cladina impexa	*Rhacomitrium canescens*
Cladina mitis	*Rhacomitrium heterostichum*
Cladina rangiferina	*Sedum spathulifolium*
Cladina stellaris	*Stereocaulon tomentosum*
Cladonia gracilis	

TABLE 14 Indicators of very dry and moderately dry soils

MOIST2	*Arctostaphylos uva-ursi*-group

Achillea lanulosa	*Homalothecium megaptilum*
Allium acuminatum	*Juniperus sibirica*
Allotropa virgata	*Lomatium dissectum*
Antennaria neglecta	*Lonicera ciliosa*
Apocynum androsaemifolium	*Lonicera hispidula*
Arbutus menziesii	*Madia madioides*
Arctostaphylos columbiana	*Mahonia aquifolium*
Arctostaphylos uva-ursi	*Peltigera aphthosa*
Bromus carinatus	*Peltigera canina*
Calamagrostis rubescens	*Peltigera membranacea*
Campanula scouleri	*Pinus ponderosa*
Carex rossii	*Pterospora andromeda*
Ceanothus sanguineus	*Quercus garryana*
Chimaphila umbellata	*Ribes lobbii*
Cladonia bellidiflora	*Ribes sanguineum*
Collinsia parviflora	*Rosa gymnocarpa*
Cryptogramma crispa	*Sanicula crassicaulis*
Cytisus scoparius	*Sanicula graveolens*
Danthonia intermedia	*Selaginella wallacei*
Danthonia spicata	*Shepherdia canadensis*
Dicranum fuscescens	*Sisyrinchium douglasii*
Dicranum tauricum	*Spiraea betulifolia*
Gaultheria ovatifolia	*Symphoricarpos hesperius*
Geranium molle	*Viola adunca*
Holodiscus discolor	*Zigadenus venenosus*

TABLE 15   Indicators of moderately dry and fresh soils

MOIST3	*Mahonia nervosa*-group

Adenocaulon bicolor	Lycopodium annotinum
Amelanchier alnifolia	Lycopodium clavatum
Arnica cordifolia	Lycopodium complanatum
Aster ciliolatus	Mahonia nervosa
Aster conspicuus	Menziesia ferruginea
Barbilophozia floerkei	Mnium spinulosum
Barbilophozia lycopodioides	Moehringia macrophylla
Brachythecium albicans	Orthilia secunda
Calypso bulbosa	Paxistima myrsinites
Camassia leichtlinii	Pedicularis racemosa
Camassia quamash	Perideridia gairdneri
Carex inops	Philadelphus lewisii
Cassiope mertensiana	Phyllodoce empetriformis
Ceanothus velutinus	Phyllodoce glanduliflora
Chimaphila menziesii	Pinus albicaulis
Clintonia uniflora	Platanthera orbiculata
Corallorhiza maculata	Polystichum lonchitis
Corallorhiza mertensiana	Potentilla glandulosa
Cornus nuttallii	Prunus virginiana
Corylus cornuta	Pyrola asarifolia
Dicranum howellii	Pyrola chlorantha
Dicranum pallidisetum	Pyrola picta
Dodecatheon hendersonii	Rhododendron albiflorum
Dodecatheon pulchellum	Ribes divaricatum
Elymus glaucus	Rosa acicularis
Erythronium oregonum	Rubus leucodermis
Fragaria vesca	Rubus ursinus
Goodyera oblongifolia	Satureja douglasii
Hieracium albiflorum	Senecio sylvaticus
Huperzia selago	Sorbus scopulina
Hypopithys lanuginosa	Sorbus sitchensis
Kindbergia oregana	Trientalis latifolia
Lathyrus nevadensis	Vaccinium membranaceum
Lathyrus ochroleucus	Vaccinium ovatum
Lilium columbianum	Vicia americana
Linnaea borealis	Viola orbiculata
Lycopodium alpinum	Viola sempervirens

TABLE 16   Indicators of fresh and very moist soils

MOIST4 *Blechnum spicant*-group	
Abies amabilis	*Dicentra formosa*
Acer circinatum	*Disporum hookeri*
Acer macrophyllum	*Disporum smithii*
Actaea rubra	*Dryopteris expansa*
Adiantum pedatum	*Dryopteris filix-mas*
Alnus sinuata - FGWT	*Equisetum hyemale* - FLOOD
Aquilegia formosa	*Equisetum telmateia* - FLOOD
Aralia nudicaulis	*Erigeron peregrinus*
Arnica latifolia	*Erythronium revolutum* - FLOOD
Aruncus dioicus	*Festuca subulata*
Asarum caudatum	*Festuca subuliflora*
Atrichum selwynii	*Galuium triflorum*
Atrichum undulatum	*Gaultheria humifusa*
Bazzania tricrenata	*Geum macrophyllum* - FGWT
Blechnum spicant	*Gymnocarpium dryopteris*
Boykinia elata - FLOOD	*Hemitomes congestum*
Calypogeia trichomanis	*Heracleum lanatum* - FGWT
Carex deweyana - FGWT	*Holcus lanatus*
Carex mertensii	*Hypericum formosum*
Cassiope stelleriana	*Isopterygium elegans*
Cassiope tetragona	*Lepidozia reptans*
Cinna latifolia - FGWT	*Listera caurina*
Circaea alpina	*Listera convallarioides*
Circaea pacifica - FGWT	*Luetkea pectinata*
Claytonia sibirica - FGWT	*Luzula parviflora*
Coptis aspleniifolia	*Lycopodium obscurum*
Cornus unalaschkensis	*Lycopodium sitchense*
Cystopteris fragilis	*Melica subulata*

TABLE 16 Indicators of fresh and very moist soils (concluded)

MOIST4	*Blechnum spicant*-group
Mitella nuda	Spiraea densiflora
Moneses uniflora	Stenanthium occidentale
Monotropa uniflora	Streptopus amplexifolius
Mycelis muralis	Streptopus roseus
Oemleria cerasiformis - FGWT	Streptopus streptopoides
Osmorhiza chilensis - FGWT	Tellima grandiflora - FGWT
Pedicularis bracteosa	Thalictrum occidentale - FGWT
Phegopteris connectilis	Tiarella laciniata
Plagiochila porelloides	Tiarella trifoliata
Plagiothecium undulatum	Tiarella unifoliata
Pogonatum alpinum	Tolmiea menziesii
Pogonatuum contortum	Trautvetteria caroliniensis - FGWT
Polystichum braunii	Trillium ovatum
Populus trichocarpa - FLOOD	Trisetum cernuum
Prenanthes alata - FLOOD	Urtica lyallii
Ranunculus uncinatus - FGWT	Vaccinium alaskaense
Rhizomnium glabrescens	Vaccinium caespitosum
Rhytidiadelphus loreus	Vaccinium deliciosum
Rosa nutkana - FGWT	Vaccinium ovalifolium
Rubus idaeus	Vahlodea atropurpurea
Rubus laciniatus	Valeriana scouleri
Rubus pedatus	Valeriana sitchensis
Rubus pubescens	Veronica americana - FLOOD
Sambucus racemosa	Viburnum edule - FLOOD
Scapania bolanderi	Viburnum trilobum
Sibbaldia procumbens	

FGWT - plants that may occur on sites with a prominently fluctuating groundwater table.
FLOOD - plants that occur on regularly flooded sites (active alluvial floodplains and stream-edge sites).

TABLE 17   Indicators of very moist and wet soils

MOIST5	***Rubus spectabilis*-group**

Agrostis aequivalvis	*Oplopanax horridus*
Athyrium filix-femina	*Parnassia fimbriata*
Calamagrostis canadensis	*Pellia neesiana*
Calamagrostis nutkaensis	*Petasites frigidus*
Caltha biflora	*Petasites palmatus* - FGWT
Caltha leptosepala	*Physocarpus capitatus* - FLOOD
Cardamine nuttallii	*Plagiomnium insigne*
Carex hendersonii - FGWT	*Ranunculus eschscholtzii* - FLOOD
Conocephalum conicum	*Ranunculus repens* - FGWT
Coptis trifolia	*Rhamnus purshianus* - FGWT
Cornus sericea - FLOOD	*Rhizomnium nudum*
Crataegus douglasii	*Ribes bracteosum* - FLOOD
Deschampsia cespitosa - FGWT	*Ribes laxiflorum*
Elymus hirsutus	*Rubus spectabilis* - FGWT
Equisetum sylvaticum	*Salix hookeriana* - FLOOD
Hippuris montana	*Sanguisorba canadensis*
Hookeria acutifolia	*Sanguisorba officinalis*
Hookeria lucens	*Saxifraga tolmiei*
Juncus effusus - FGWT	*Senecio triangularis*
Juncus ensifolius - FGWT	*Sphagnum girgensohnii* - FGWT
Kindbergia praelonga - FGWT	*Spiraea douglasii* - FGWT
Leptarrhena pyrolifolia	*Spiraea menziesii* - FGWT
Leucolepis menziesii	*Stachys cooleyae* - FLOOD
Loiseleuria procumbens	*Stachys mexicana* - FLOOD
Lonicera involucrata - FLOOD	*Stellaria calycantha*
Maianthemum dilatatum - FLOOD	*Stellaria crispa*
Marchantia polymorpha	*Vaccinium uliginosum*
Mitella breweri	*Veratrum eschscholtzii*
Mitella ovalis	*Viola glabella* - FLOOD
Mitella pentandra	

FGWT -   plants that may occur on sites with prominently fluctuating groundwater table.
FLOOD -   plants that occur on regularly flooded sites (active alluvial floodplains and stream-edge sites).

TABLE 18   Indicators of wet and very wet soils

MOIST6	*Lysichitum americanum*-group

Andromeda polifolia	*Nuphar polysepalum*
Angelica genuflexa	*Oenanthe sarmentosa*
Aulacomnium palustre	*Philonotis fontana*
Cardamine breweri	*Platanthera dilatata*
Carex anthoxanthea	*Rhizomnium magnifolium*
Carex laeviculmis	*Rhynchospora alba*
Carex livida	*Scirpus microcarpus*
Carex obnupta	*Siphula ceratites*
Carex sitchensis	*Sphagnum capillifolium*
Drosera rotundifolia	*Sphagnum fallax*
Eriophorum angustifolium	*Sphagnum fuscum*
Fauria crista-galli	*Sphagnum papillosum*
Gentiana douglasiana	*Sphagnum tenellum*
Gentiana sceptrum	*Tofieldia glutinosa*
Kalmia occidentalis	*Torreyochloa pauciflora*
Ledum groenlandicum	*Trichophorum cespitosum*
Lysichitum americanum	*Trientalis arctica*
Malus fusca	*Vaccinium oxycoccos*
Menyanthes trifoliata	*Viola palustris*
Myrica gale	

Soil Nutrients

Soil nutrient regime (SNR) is the average amount of essential soil nutrients that are available to vascular plants over several years. Krajina (1969) adopted six classes (A to F) of SNRs (so-called relative SNRs) and applied them in different climates and for soils with different SMRs. For example, the relatively nutrient-poorest soil within any SMR and climate was considered oligotrophic (A) and the nutrient-richest [excluding the soil with an excessive content of bases - hypereutrophic (F)] was eutrophic (E). Despite ease in field identification (see Green et al. 1984), relative SMRs express inconsistently the actual level of available soil nutrients, such that an oligotrophic SNR in one situation could be poorer or richer than in another situation.

To facilitate the use of plants as indicators of soil nutrients, we used actual SNRs to characterize plant-nutrient relationships. Recently, Courtin et al. (1987) proposed a tentative classification of SNRs and correlated the delineated SNRs to vegetation and forest productivity. Kabzems and Klinka (1987) concurred but showed that the content of rooting-zone mineralizable nitrogen (kg/ha) can be used as the single differentiating characteristic (cf. Zöttl 1960; Bremner 1965; Shumway and Atkinson 1978; Powers 1980; Smith et al. 1981). In consequence, the adopted tentative SNR classification delineates major segments of the available soil nitrogen gradient (Table 19) - the property suggested by Ballard (Dept. of Soil Science, Univ. of British Columbia; pers. comm.) as the most useful differentiating characteristic.

It is now generally accepted that nitrogen is one of the most important, and generally most deficient, soil nutrient factor limiting the growth of plants (Black 1968; Tisdale and Nelson 1975; Armson 1977; Pritchett 1979; Ballard and Carter 1986). Due to the complexity of plant nutrient needs, a single differentiating characteristic, even if representing the most important nutrient element, cannot provide for a complete SNR classification. Further developments in the classification should consider other nutrients as differentiating characteristics for classes of lower rank.

Actual SNRs were identified by extrapolating mineralizable nitrogen data (Klinka, Feller, and Lowe 1981, 1982; Roy 1984; Kabzems 1985; Carter and Klinka 1987) and, in areas where these data were not applicable, by interpreting the data on forest floor pH and C/N ratio, soil total N (kg/ha) and sum of available calcium, magnesium, and potassium (kg/ha) (Table 19). This information was used to stratify vegetation data and calibrate indicator species of available soil nitrogen.

TABLE 19   Means and standard deviations (in parentheses) of soil properties used to characterize actual soil nutrient regimes in coastal British Columbia [from Courtin et al. (1987) and Kabzems and Klinka (1987)]

Property	Soil nutrient regime				
	Very poor (VP)	Poor (P)	Medium (M)	Rich (R)	Very rich (VR)
Mineralizable-nitrogen (kg/ha)	<10[1]	18 (7)	54 (20)	113 (39)	242 (130)
Forest floor pH	3.8 (0.25)	4.0 (0.47)	4.1 (0.62)	4.5 (0.43)	5.0 (0.43)
Forest floor C/N ratio	73 (7.4)	42 (7.3)	34 (6.8)	20 (4.5)	21 (4.5)
Total nitrogen (kg/ha)	1743 (1786)	2328 (490)	3193 (865)	4108 (2281)	7121 (2174)
Sum of available calcium, magnesium, and potassium (kg/ha)	1386 (1683)	873 (650)	1225 (577)	1743 (1088)	5066 (1961)

[1] Estimated value.

Indicators of Soil Nitrogen

As is the case with soil moisture, precise and site-specific information on available nitrogen in forest soils is not available. To avoid expensive sampling and chemical analysis, plant indicators of soil nitrogen are an effective substitute for quantitative data in site diagnosis. Many workers have concluded that the natural vegetation produces the best measure of available nitrogen levels in the soil.

There are 378 indicator species in three ISGs for indication of soil nitrogen in coastal British Columbia (Tables 21 through 23). Based on the indicator values adopted for diagnosing soil nitrogen (Table 20), these groups represent wide and overlapping segments of the soil nitrogen gradient.

With a few exceptions (see Appendix II) the species included in the *Gaultheria shallon*-group (Table 21) inhabit acid substrates (organic or mineral materials with pH values approximately <4.5), and have been designated by Dahl (1956) and Krajina et al. (1986) as oxylophytic. The species included in the *Tiarella trifoliata*-group (Table 23) inhabit substrates that contain easily available nitrogen as a result of strong nitrification. Krajina et al. (1986) designated such species as nitrophytic as most of them have been reported to accumulate nitrate-nitrogen in their foliage (Hesselman 1917; Sillinger 1939; Ellenberg 1952; Mezera 1957; Krajina et al. 1986). Other comparative and experimental studies have demonstrated definite floristic differences between plant communities on sites with highly contrasting levels of available soil nitrogen (Olsen 1921; de Coulon 1923; Gessner 1932; Rees and Sidrak 1956; Bradshaw et al. 1964; Piggot and Taylor 1964). Thus, the division into oxylophytic and nitrophytic species probably reflects basic ecological differences between plant species in response to nitrogen supply.

The species that usually inhabit calcium-rich substrates or may grow on alkaline and saline substrates, and thus may be used as indicators of these edaphic conditions, are marked in Table 23.

TABLE 20   Indicators of soil nitrogen: synopsis of indicator species groups

Symbol	Name of ISG	Range of actual SNRs -
NITR1	*Gaultheria shallon*-group	very poor to poor (-medium) (P)
NITR2	*Tiarella unifoliata*-group	(poor-) medium (-rich) (M)
NITR3	*Tiarella trifoliata*-group	rich to very rich (R)

TABLE 21   Indicators of nitrogen-poor soils

NITR1   *Gaultheria shallon*-group

Aira praecox	*Cassiope mertensiana*
Allotropa virgata	*Cassiope stelleriana*
Andromeda polifolia	*Chimaphila umbellata*
Arctostaphylos columbiana	*Cladina arbuscula*
Arctostaphylos uva-ursi	*Cladina impexa*
Barbilophozia floerkei	*Cladina mitis*
Barbilophozia lycopodioides	*Cladina rangiferina*
Bazzania tricrenata	*Cladina stellaris*
Blechnum spicant	*Cladonia bellidiflora*
Boschniakia hookeri	*Cladonia gracilis*
Calypogeia trichomanis	*Cladothamnus pyroliflorus*
Campanula scouleri	*Clintonia uniflora*

TABLE 21   Indicators of nitrogen-poor soils (concluded)

NITR1	*Gaultheria shallon*-group

Coptis aspleniifolia	*Paxistima myrsinites*
Coptis trifolia	*Peltigera aphthosa*
Corallorhiza maculata	*Peltigera canina*
Corallorhiza mertensiana	*Peltigera membranacea*
Cornus canadensis	*Phyllodoce empetriformis*
Cornus unalaschkensis	*Plagiothecium undulatum*
Cryptogramma crispa	*Platanthera orbiculata*
Danthonia intermedia	*Pleurozium schreberi*
Danthonia spicata	*Polytrichum piliferum*
Dicranum fuscescens	*Ptilium crista-castrensis*
Dicranum howellii	*Rhacomitrium canescens*
Dicranum pallidisetum	*Rhacomitrium heterostichum*
Dicranum tauricum	*Rhododendron albiflorum*
Drosera rotundifolia	*Rhynchospora alba*
Empetrum nigrum	*Rhytidiadelphus loreus*
Equisetum sylvaticum	*Rhytidiopsis robusta*
Eriophorum angustifolium	*Rubus pedatus*
Fauria crista-galli	*Scapania bolanderi*
Festuca occidentalis	*Sedum spathulifolium*
Gaultheria humifusa	*Selaginella wallacei*
Gaultheria ovatifolia	*Sibbaldia procumbens*
Gaultheria shallon	*Siphula ceratites*
Gentiana douglasiana	*Sorbus sitchensis*
Geocaulon lividum	*Sphagnum fallax*
Goodyera oblongifolia	*Sphagnum fuscum*
Homalothecium megaptilum	*Sphagnum girgensohnii*
Hookeria acutifolia	*Sphagnum nemoreum*
Hookeria lucens	*Sphagnum papillosum*
Huperzia selago	*Sphagnum tenellum*
Hylocomium splendens	*Stereocaulon tomentosum*
Hypopithys lanuginosa	*Streptopus streptopoides*
Isopterygium elegans	*Trichophorum caespitosum*
Kalmia occidentalis	*Trientalis arctica*
Ledum groenlandicum	*Vaccinium alaskaense*
Lepidozia reptans	*Vaccinium caespitosum*
Listera cordata	*Vaccinium deliciosum*
Loiseleuria procumbens	*Vaccinium membranaceum*
Luzula multiflora	*Vaccinium myrtilloides*
Lycopodium clavatum	*Vaccinium ovalifolium*
Lycopodium complanatum	*Vaccinium ovatum*
Lycopodium obscurum	*Vaccinium oxycoccos*
Lycopodium sitchense	*Vaccinium parvifolium*
Menziesia ferruginea	*Vaccinium scoparium*
Orthilia secunda	*Vaccinium uliginosum*

TABLE 22   Indicators of nitrogen-medium soils

NITR2	*Tiarella unifoliata*-group

Achillea lanulosa	Eriophyllum lanatum
Agrostis aequivalvis	Erythronium oregonum
Allium acuminatum	Fragaria vesca
Allium cernuum	Fragaria virginiana - CALC
Amelanchier alnifolia	Gentiana sceptrum
Apocynum androsaemifolium	Geranium molle
Arnica cordifolia	Hemitomes congestum
Arnica latifolia	Hippuris montana
Aster ciliolatus	Holcus lanatus
Aulacomnium palustre	Holodiscus discolor
Brachythecium albicans	Hypericum formosum
Bromus carinatus	Juncus effusus
Calamagrostis canadensis	Juncus ensifolius
Calamagrostis nutkaensis - SALI	Juniperus scopulorum - ALKA
Calamagrostis rubescens	Juniperus sibirica
Calypso bulbosa	Leptarrhena pyrolifolia
Carex anthoxanthea	Lilium columbianum
Carex inops	Listera caurina
Carex laeviculmis	Lonicera ciliosa
Carex livida	Lonicera hispidula
Carex rossii	Lonicera utahensis
Cassiope tetragona	Luetkea pectinata
Ceanothus sanguineus	Luzula parviflora
Ceanothus velutinus	Lycopodium alpinum
Chimaphila menziesii	Lycopodium annotinum
Circaea alpina	Madia madioides
Collinsia parviflora	Mahonia aquifolium
Cystopteris fragilis	Mahonia nervosa
Cytisus scoparius	Menyanthes trifoliata
Dryopteris expansa	Mitella nuda
Equisetum arvense	Mnium spinulosum

TABLE 22   Indicators of nitrogen-medium soils (concluded)

NITR2	*Tiarella unifoliata*-group

Moneses uniflora	Ribes lobbii
Monotropa uniflora	Ribes sanguineum
Montia parvifolia	Rosa acicularis
Myrica gale	Rosa gymnocarpa
Pedicularis bracteosa	Rubus ursinus
Pedicularis racemosa	Salix bebbiana
Perideridia gairdneri	Salix hookeriana
Philadelphus lewisii	Salix scouleriana
Philonotis fontana	Salix sitchensis
Phyllodoce glanduliflora	Sanguisorba officinalis
Pinus albicaulis	Shepherdia canadensis
Pinus ponderosa - CALC	Sisyrinchium douglasii
Plagiochila porelloides - CALC	Sorbus scopulina
Platanthera dilatata	Spiraea betulifolia
Pogonatum alpinum	Spiraea densiflora
Pogonatum contortum	Spiraea douglasii
Polystichum lonchitis	Symphoricarpos hesperius
Potentilla glandulosa	Tiarella unifoliata
Pterospora andromeda	Timmia austriaca
Pyrola asarifolia	Trientalis latifolia
Pyrola chlorantha	Vahlodea atropurpurea
Pyrola picta	Vicia americana
Ranunculus occidentalis	Viola adunca
Rhizomnium glabrescens	Viola orbiculata
Rhizomnium nudum	Viola sempervirens
Rhytidiadelphus triquetrus	Zigadenus venenosus
Ribes divaricatum	

ALKA   - a plant that may occur on alkaline soils.
CALC   - plants that occur on calcium-rich soils.
SALI   - a plant that may occur on saline soils.

TABLE 23   Indicators of nitrogen-rich soils

NITR3	*Tiarella trifoliata*-group

Acer circinatum	*Dodecatheon hendersonii*
Acer glabrum	*Dodecatheon pulchellum* - ALKA
Acer macrophyllum	*Dryopteris filix-mas*
Achlys triphylla	*Elymus glaucus*
Actaea rubra - NITR	*Elymus hirsutus*
Adenocaulon bicolor - NITR	*Epilobium angustifolium* - NITR
Adiantum pedatum - CALC	*Epilobium latifolium* - NITR
Agropyron spicatum - ALKA, NITR	*Equisetum hyemale* - CALC
Alnus rubra	*Equisetum telmateia* - NITR
Alnus sinuata	*Erigeron peregrinus*
Angelica genuflexa - NITR	*Erythronium revolutum*
Aquilegia formosa - NITR	*Festuca subulata*
Aralia nudicaulis - NITR	*Festuca subuliflora* - NITR
Aruncus dioicus - NITR	*Fritillaria lanceolata*
Asarum caudatum - NITR	*Galium aparine* - NITR
Aster conspicuus - ALKA, NITR	*Galium triflorum* - NITR
Athyrium filix-femina - NITR	*Geum macrophyllum* - NITR
Atrichum selwynii	*Gymnocarpium dryopteris*
Atrichum undulatum	*Heracleum lanatum* - NITR
Boykinia elata - NITR	*Heuchera micrantha*
Bromus vulgaris - NITR	*Kindbergia praelonga*
Caltha biflora	*Lathyrus nevadensis*
Caltha leptosepala	*Lathyrus ochroleucus*
Camassia leichtlinii	*Leucolepis menziesii*
Camassia quamash	*Listera convallarioides* - NITR
Cardamine breweri - NITR	*Lomatium dissectum* - NITR
Cardamine nuttallii - NITR	*Lonicera involucrata* - NITR
Carex deweyana - NITR	*Lupinus arcticus*
Carex hendersonii - NITR	*Lupinus nootkatensis*
Carex mertensii - NITR	*Lysichitum americanum* - NITR
Carex obnupta - NITR	*Maianthemum dilatatum*
Carex sitchensis	*Malus fusca*
Cinna latifolia - NITR	*Melica subulata* - NITR
Circaea pacifica - NITR	*Mitella breweri* - NITR
Claytonia sibirica - NITR	*Mitella ovalis* - NITR
Conocephalum conicum - CALC	*Mitella pentandra* - NITR
Cornus nuttallii	*Moehringia macrophylla* - NITR
Cornus sericea	*Mycelis muralis* - NITR
Corylus cornuta - CALC	*Oemleria cerasiformis*
Crataegus douglasii	*Oenanthe sarmentosa* - NITR
Deschampsia caespitosa - CALC, SALI	*Oplopanax horridus* - NITR
Dicentra formosa - NITR	*Osmorhiza chilensis* - NITR
Disporum hookeri - NITR	*Parnassia fimbriata* - CALC
Disporum smithii	*Pellia neesiana*
Disporum trachycarpum	*Petasites frigidus* - NITR

TABLE 23   Indicators of nitrogen-rich soils (concluded)

NITR3	*Tiarella trifoliata*-group

Petasites palmatus - NITR	*Senecio sylvaticus* - NITR
Phegopteris connectilis - CALC, NITR	*Senecio triangularis* - NITR
Physocarpus capitatus	*Senecio vulgaris* - NITR
Picea sitchensis - SALI	*Smilacina racemosa* - NITR
Plagiomnium insigne	*Smilacina stellata* - NITR
Polystichum braunii - NITR	*Spiraea menziesii*
Polystichum munitum - NITR	*Stachys cooleyae* - NITR
Populus tremuloides	*Stachys mexicana* - NITR
Populus trichocarpa - NITR	*Stellaria crispa* - NITR
Prenanthes alata - NITR	*Stenanthium occidentale*
Prunus virginiana - ALKA, NITR	*Streptopus amplexifolius* - NITR
Ranunculus eschscholtzii	*Streptopus roseus* - NITR
Ranunculus repens - NITR	*Symphoricarpos albus*
Ranunculus uncinatus - NITR	*Tellima grandiflora* - NITR
Rhamnus purshianus	*Thalictrum occidentale* - NITR
Rhizomnium magnifolium	*Tiarella laciniata* - NITR
Ribes bracteosum - NITR	*Tiarella trifoliata* - NITR
Ribes lacustre - NITR	*Tolmiea menziesii* - NITR
Ribes laxiflorum - NITR	*Torreyochloa pauciflora* - NITR
Rosa nutkana	*Trautvetteria caroliniensis* - NITR
Rubus idaeus - NITR	*Trillium ovatum* - NITR
Rubus laciniatus	*Trisetum cernuum* - NITR
Rubus leucodermis - NITR	*Urtica lyallii* - NITR
Rubus parviflorus - NITR	*Valeriana scouleri* - NITR
Rubus pubescens - NITR	*Valeriana sitchensis* - NITR
Rubus spectabilis - NITR	*Veratrum eschscholtzii* - NITR
Sambucus racemosa - NITR	*Veronica americana* - NITR
Sanicula crassicaulis - NITR	*Viburnum edule* - NITR
Sanicula graveolens - NITR	*Viburnum trilobum* - NITR
Satureja douglasii - NITR	*Viola glabella* - NITR
Scirpus microcarpus - NITR	

ALKA - plants that may occur on alkaline soils.
CALC - plants that occur on calcium-rich soils.
NITR - nitrophytic species
SALI - plant that may occur on saline soils.

Ground Surface Materials

Forest understory vegetation obtains water and nutrients from the surface soil layer. Thus the material of this layer will likely influence the composition of understory vegetation and the potential for the establishment and growth of plants. We have recognized five types of materials that may occur in various combinations and with varying coverage on a forest site to describe relationships between plants and ground surface materials (Table 24).

The forest floor is defined as all dead vegetative and organic matter, including litter and incorporated humus, on the mineral soil surface under forest vegetation (Canada Depart. of Agric. 1976). The forest floor may be very thin (<1 cm) and discontinuous or very thick (>40 cm). It may contain large amounts of decaying wood and it may be abruptly separated from the mineral soil or underlain by an Ah horizon rich in organic material.

Six types of surface soil organic horizons may be distinguished in the forest floor (Canada Soil Survey Committee 1978). They are identified according to soil moisture regime and degree of decomposition: well aerated (terrestrial or upland) L (litter), F (fermented), and H (humus) horizons; and poorly aerated (semi-terrestrial or wetland) Of (fibric), Om (mesic), and Oh (humic) horizons. A humus form is a combination and expression of these organic horizons (including the mineral Ah horizon) and is the most biologically active component of the soil. The forest floor releases nutrients through mineralization of organic materials and consequently, humus form has been one of the most useful indices of available nutrients (Kubiëna 1953; Wilde 1954, 1966, 1971). Humus form and ground surface materials data were used to stratify vegetation data and to calibrate the indicator species. More detailed information on the classification and identification of humus forms of British Columbia can be found in Klinka, Green, Trowbridge, and Lowe (1981).

TABLE 24   Principal types of ground surface materials

Ground surface material	Differentiating characteristic
Mor humus forms (MOR)	a layer approximately 5 cm (or more) of compacted (matted by fungal mycelia) organic materials (of more than 17% organic carbon), overlying mineral soil
Moder and Mull humus forms (MDML)	a layer less than 5 cm of friable (partly fragmented or comminuted by soil fauna) organic materials (of more than 17% organic carbon), overlying mineral soil (an Ah horizon in Mulls)
Mineral soil (MS)	unconsolidated mineral materials with less than 17% organic carbon and particle size less than 2 mm in diameter; often temporarily covered by litter
Very shallow soil layer over coarse fragments and bedrock (VSS)	unconsolidated mineral materials greater than 2 mm in diameter; gravel, stones, boulders, and bedrock
Surface water (SW)	free water at or above the ground surface

Indicators of Ground Surface Materials

The ground surface materials present on a site determine much of the potential for establishment and growth of a forest. Thus indicators of the dominant type(s) of ground surface materials on a site can be useful in inferring humus forms, evaluating changes in ground cover following natural disturbances, harvesting, site preparation, and other treatments affecting the ground surface, and in predicting vegetation development following disturbance of the tree layers and/or surface soil layers. There are 289 indicator species in five ISGs for indication of ground surface materials in coastal British Columbia (Tables 25 through 30).

Indicators of compacted forest floors or Mors (the *Vaccinium parvifolium-* group; Table 26) are all oxylophytic species (cf. Table 21), while indicators of friable forest floors or Moders and Mulls (the *Polystichum munitum*-group; Table 27) are indicators of nitrogen-rich soils (see Table 23). Indicators of exposed mineral soil (the *Anaphalis margaritacea*-group; Table 28) include predominantly shade-intolerant species that usually inhabit temporarily or regularly disturbed sites. Indicators of surface water (the *Sphagnum*-group; Table 30) are all obligatory indicators of wet soils (see Table 18).

From the forest manager's viewpoint, vegetation changes during succession are either positive or negative and influence the way he chooses to disturb forest floor materials. For example, the persistent occurrence of *Gaultheria shallon* on cut-over and burnt sites is indicative of little disturbance and virtually no change in the decomposition rate of acid and compacted forest floors (Mor humus forms). In contrast, *Polystichum munitum* indicates friable forest floors (Moder and Mull humus forms).

TABLE 25   Indicators of ground surface materials - synopsis of indicator species groups

Symbol	Name of ISG	Type of ground surface material
GSM1	*Vaccinium parvifolium*-group	Mor humus forms (MOR)
GSM2	*Polystichum munitum*-group	Moder and Mull humus forms (MDML)
GSM3	*Anaphalis margaritacea*-group	exposed mineral soils (EMS)
GSM4	*Selaginella wallacei*-group	very shallow soil layer over exposed coarse fragments and bedrock (VSS)
GSM5	*Sphagnum*-group	surface groundwater table (SW) for most of the year except the driest months of summer

TABLE 26   Indicators of Mor humus forms

GSM1	*Vaccinium parvifolium*-group

Allotropa virgata	*Listera cordata*
Arctostaphylos uva-ursi	*Loiseleuria procumbens*
Barbilophozia floerkei	*Lycopodium clavatum*
Barbilophozia lycopodioides	*Lycopodium complanatum*
Bazzania tricrenata	*Lycopodium obscurum*
Blechnum spicant	*Lycopodium sitchense*
Boschniakia hookeri	*Menziesia ferruginea*
Calypogeia trichomanis	*Orthilia secunda*
Campanula scouleri	*Paxistima myrsinites*
Cassiope mertensiana	*Peltigera aphthosa*
Cassiope stelleriana	*Peltigera canina*
Chimaphila umbellata	*Peltigera membranacea*
Cladothamnus pyroliflorus	*Phyllodoce empetriformis*
Clintonia uniflora	*Plagiothecium undulatum*
Coptis aspleniifolia	*Platanthera orbiculata*
Coptis trifolia	*Pleurozium schreberi*
Corallorhiza maculata	*Ptilium crista-castrensis*
Corallorhiza mertensiana	*Rhododendron albiflorum*
Cornus canadensis	*Rhytidiadelphus loreus*
Cornus unalaschkensis	*Rhytidiopsis robusta*
Dicranum fuscescens	*Rubus pedatus*
Dicranum howellii	*Scapania bolanderi*
Dicranum pallidisetum	*Sibbaldia procumbens*
Dicranum tauricum	*Siphula ceratites*
Empetrum nigrum	*Sorbus sitchensis*
Equisetum sylvaticum	*Sphagnum girgensohnii*
Festuca occidentalis	*Streptopus streptopoides*
Gaultheria humifusa	*Tsuga heterophylla* [1]
Gaultheria ovatifolia	*Tsuga mertensiana* [1]
Gaultheria shallon	*Vaccinium alaskaense*
Geocaulon lividum	*Vaccinium caespitosum*
Goodyera oblongifolia	*Vaccinium deliciosum*
Homalothecium megaptilum	*Vaccinium membranaceum*
Hookeria acutifolia	*Vaccinium myrtilloides*
Hookeria lucens	*Vaccinium ovalifolium*
Huperzia selago	*Vaccinium ovatum*
Hylocomium splendens	*Vaccinium parvifolium*
Hypopithys lanuginosa	*Vaccinium scoparium*
Lepidozia reptans	*Vaccinium uliginosum*

[1] Natural regeneration.

TABLE 27   Indicators of Moder and Mull humus forms

GSM2   *Polystichum munitum*-group

Acer circinatum	Elymus hirsutus
Acer glabrum	Erigeron peregrinus
Acer macrophyllum	Festuca subulata
Achlys triphylla	Festuca subuliflora
Actaea rubra	Fritillaria lanceolata
Adenocaulon bicolor	Galium aparine - DIST
Adiantum pedatum	Galium triflorum
Agropyron spicatum	Gymnocarpium dryopteris
Alnus rubra - DIST	Heracleum lanatum
Alnus sinuata - DIST	Heuchera micrantha
Aquilegia formosa - DIST	Kindbergia praelonga
Aralia nudicaulis	Lathyrus nevadensis
Aruncus dioicus - DIST	Lathyrus ochroleucus
Asarum caudatum	Listera convallarioides
Aster conspicuus	Lomatium dissectum
Athyrium filix-femina	Lonicera involucrata
Boykinia elata	Maianthemum dilatatum
Bromus vulgaris	Melica subulata
Caltha biflora	Mitella breweri
Caltha leptosepala	Mitella ovalis
Camassia leichtlinii	Mitella pentandra
Camassia quamash	Moehringia macrophylla
Cardamine nuttallii	Mycelis muralis - DIST·
Carex deweyana	Oemleria cerasiformis
Carex hendersonii	Oplopanax horridus
Cinna latifolia	Osmorhiza chilensis
Circaea pacifica	Parnassia fimbriata
Claytonia sibirica	Pellia neesiana
Cornus nuttallii	Phegopteris connectilis
Cornus sericea	Physocarpus capitatus
Corylus cornuta	Polystichum braunii
Crataegus douglasii	Polystichum munitum
Dicentra formosa - DIST	Populus tremuloides
Disporum hookeri	Populus trichocarpa
Disporum smithii	Prenanthes alata
Disporum trachycarpum	Prunus virginiana
Dodecatheon hendersonii	Ranunculus eschscholtzii
Dodecatheon pulchellum	Rhamnus purshianus
Dryopteris filix-mas	Ribes bracteosum
Elymus glaucus	Ribes lacustre
	Ribes laxiflorum

Table 27   Indicators of Moder and Mull humus forms (concluded)

GSM2	Polystichum munitum-group

Rosa nutkana	*Symphoricarpos albus*
Rubus parviflorus	*Tellima grandiflora*
Rubus pubescens	*Thalictrum occidentale*
Rubus spectabilis	*Tiarella laciniata*
Sambucus racemosa - DIST	*Tiarella trifoliata*
Sanicula crassicaulis	*Tolmiea menziesii*
Sanicula graveolens	*Trautvetteria caroliniensis*
Satureja douglasii	*Trillium ovatum*
Senecio triangularis	*Trisetum cernuum*
Smilacina racemosa	*Valeriana scouleri*
Smilacina stellata	*Valeriana sitchensis*
Stachys cooleyae	*Veratrum eschscholtzii*
Stachys mexicana	*Veronica americana*
Stellaria crispa	*Viburnum edule*
Stenanthium occidentale	*Viburnum trilobum*
Streptopus amplexifolius	*Viola glabella*
Streptopus roseus	

DIST - plants that occur on temporarily or regularly disturbed sites.

TABLE 28   Indicators of exposed mineral soil

GSM3	*Anaphalis margaritacea*-group

Achillea lanulosa	*Holcus lanatus*
Anaphalis margaritacea	*Hypochaeris radicata*
Apocynum androsaemifolium	*Leucolepis menziesii*
Atrichum selwynii	*Madia madioides*
Atrichum undulatum	*Marchantia polymorpha*
Carex mertensii	*Plagiomnium insigne*
Cytisus scoparius	*Pogonatum contortum*
Equisetum hyemale	*Ranunculus repens*
Equisetum telmateia	*Ranunculus uncinatus*
Fragaria vesca	*Saxifraga tolmiei*
Fragaria virginiana	*Senecio sylvaticus*
Geum macrophyllum	*Senecio vulgaris*
Hieracium albiflorum	

TABLE 29   Indicators of very shallow soils

GSM4 *Selaginella wallacei*-group	
Aira caryophyllea	Eriophyllum lanatum
Aira praecox	Geranium molle
Asplenium trichomanes	Juniperus scopulorum
Bartramia pomiformis	Luzula multiflora
Carex rossii	Montia parvifolia
Cladina arbuscula	Pilophoron clavatus
Cladina impexa	Polypodium glycyrrhiza
Cladina mitis	Polypodium scouleri
Cladina rangiferina	Polytrichum piliferum
Cladina stellaris	Rhacomitrium canescens
Cladonia bellidiflora	Rhacomitrium heterostichum
Cladonia gracilis	Saxifraga ferruginea
Claopodium crispifolium	Sedum spathulifolium
Collinsia parviflora	Selaginella wallacei
Cryptogramma crispa	Sisyrinchium douglasii
Danthonia intermedia	Stereocaulon tomentosum
Danthonia spicata	Zigadenus venenosus

TABLE 30   Indicators of surface groundwater table

GSM5 *Sphagnum-group*	
Andromeda polifolia	Myrica gale
Angelica genuflexa	Nuphar polysepalum
Aulacomnium palustre	Oenanthe sarmentosa
Cardamine breweri	Philonotis fontana
Carex anthoxanthea	Platanthera dilatata
Carex laeviculmis	Rhizomnium magnifolium
Carex livida	Rhynchospora alba
Carex obnupta	Scirpus microcarpus
Carex sitchensis	Sphagnum capillifolium
Drosera rotundifolia	Sphagnum fallax
Eriophorum angustifolium	Sphagnum fuscum
Fauria crista-galli	Sphagnum papillosum
Gentiana douglasiana	Sphagnum tenellum
Gentiana sceptrum	Tofieldia glutinosa
Kalmia occidentalis	Torreyochloa pauciflora
Ledum groenlandicum	Trichophorum cespitosum
Lysichitum americanum	Trientalis arctica
Malus fusca	Vaccinium oxycoccos
Menyanthes trifoliata	Viola palustris

5

INDICATOR PLANT ANALYSIS

Site diagnosis is usually confined to estimates of soil moisture regimes (SMRs) and soil nutrient regimes (SNRs), providing that regional climate can be identified from biogeoclimatic maps. Thus, this chapter will demonstrate the application of indicator plant analysis in soil moisture and soil nutrient diagnosis, but the analysis can be extended to types of climate and ground surface materials. Indicator plant analysis should be always accompanied by environmental analysis as described by Klinka et al. (1984).

To obtain the most precise results it is necessary to sample the vegetation of a site, recording all species and their cover. A sample plot of approximately 20 m x 20 m is located as to represent an individual ecosystem; an area that is relatively uniform in vegetation and soils. Plot size can be reduced when sampling non-forested vegetation (grassland, heath, or meadow communities) or particular microsites (decaying wood, mounds, depressions, or disturbed surfaces). All plant species or only those occurring in the shrub, herb, and moss strata (understory vegetation) present in a sample plot are identified and their percent cover estimated. Percent cover refers to the proportion of the plot that is covered by a vertical projection of the crown/foliage onto the ground surface of all individuals of a species. Species cover is estimated using either a six-class (Table 31) or the ten-class scale (Table 32) that is used in this guide. Comparison charts, to assist in estimating percent cover, are given in Figure 5.

Individual-species, spectral, and index methods are described, demonstrated, and evaluated using the vegetation of four test plots to diagnose their actual SMRs and SNRs (see Tables 33 and 34). The choice of an appropriate method will depend on the vegetation under examination and importance of the analysis. The floristic composition of a plant community depends on site quality, the community itself, and site history. Some communities are species-poor, such as fireweed-dominated communities on recently disturbed site, but the majority are species-diverse, with species cover and distribution varying with the community and microsites present (canopy openings, microtopography, and ground surface materials). Rowe (1956) suggested that an adequate vegetation sample for indicator plant analysis should contain more than 12 species.

TABLE 31   Six-class cover scale (from Klinka et al. 1984)

Code	Midpoint of cover class (%)	Cover class interval (%)		
+	0.5			<1
1	7.0	1.	–	5
2	15.0	5.1	–	25
3	37.5	25.1	–	50
4	62.5	50.1	–	75
5	87.5			>75

TABLE 32   Domin-Krajina's ten-class cover scale (from Brooke et al. 1970)

Code	Midpoint of cover class (%)	Cover class interval (%)		
+	0.2	0.1	–	0.3
1	0.7	0.4	–	1.0
2	1.6	1.1	–	2.2
3	3.6	2.3	–	5.0
4	7.5	5.1	–	10.0
5	15.0	10.1	–	20.0
6	26.5	20.1	–	33.0
7	41.5	33.1	–	50.0
8	60.0	50.1	–	70.0
9	85.0	70.1	–	100.0

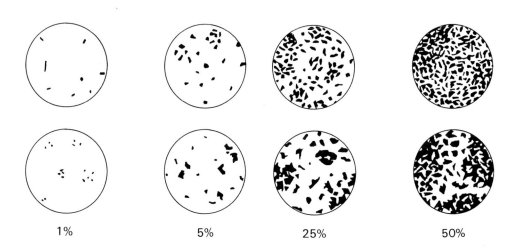

1%   5%   25%   50%

FIGURE 5   Comparison charts for visual estimation of foliage cover (from Walmsley et al. 1980)

Those who are thoroughly familiar with the vegetation, indicative values of plants, and sites of coastal British Columbia are often able to estimate site quality with considerable accuracy from a very brief examination. Unfortunately, this approach is not possible for everyone, and more objective methods that can be consistently used by all are desirable.

Indicator plant analysis is still under development and requires testing. While it is not difficult to rank a series of sites according to a given site attribute, such as soil moisture, difficulties are encountered when inferring the actual value of that attribute. For example, our data analysis suggests that the slightly dry and fresh SMRs or rich and very rich SNRs cannot be distinguished by indicator plants. Until the results of indicator plant analysis of many environmentally well characterized plant communities are calibrated in relation to the scales used to measure site attributes, the results of indicator plant analysis should be regarded as approximations.

TABLE 33   Vegetation of test plots used to demonstrate the methods of indicator plant analysis

Test plot	Saanich Peninsula	Cowichan Lake	MacMillan Park	Manning Park
Species	Species cover code/midpoint percent cover			
Abies grandis			4/7.5	
Abies lasiocarpa				3/3.6
Acer macrophyllum		+/0.2		
Achlys triphylla	1/0.7	1/0.7	7/41.5	
Adenocaulon bicolor			2/1.6	
Adiantum pedatum		+/0.2		
Alnus rubra		6/26.5		
Amelanchier alnifolia				1/0.7
Arctostaphylos uva-ursi				2/1.6
Athyrium filix-femina		7/41.5	1/0.7	
Blechnum spicant		1/0.7		
Bromus sitchensis		+/0.2		
Bromus vulgaris	1/0.7	1/0.7		
Calamagrostis rubescens				4/7.5
Carex deweyana		1/0.7		
Carex hendersonii			3/3.6	
Circaea alpina		+/0.2		
Cladonia apocodarpa				6/26.5
Cladonia gracilis				4/7.5
Cladonia pyxidata				1/0.7
Claytonia sibirica		+/0.2		
Conocephalum conicum		+/0.2		
Corallorhiza maculata	+/0.2			
Cornus unalaschkensis	3/3.6			
Cornus sericea	1/0.7			
Disporum hookeri		+/0.2	3/3.6	
Dryopteris expansa			2/1.6	

TABLE 33   Vegetation of test plots used to demonstrate the methods of indicator plant analysis (continued)

Test plot Species	Saanich Peninsula	Cowichan Lake	MacMillan Park	Manning Park
	Species cover code/midpoint percent cover			
Equisetum arvense		+/0.2		
Equisetum telmateia		+/0.2		
Festuca occidentalis	+/0.2			
Festuca subulata	1/0.7			
Galium triflorum	1/0.7	+/0.2	6/26.5	
Gaultheria shallon	5/15.0			
Glyceria elata		+/0.2		
Hieracium albiflorum				1/0.7
Hylocomium splendens			4/7.5	
Kindbergia oregana	5/15.0	+/0.2	4/7.5	
Kindbergia praelonga		3/3.6		
Lathyrus nevadensis	1/0.7			
Leucolepis menziesii		3/3.6		
Lonicera ciliosa	1/0.7			
Lonicera hispidula	+/0.2			
Lupinus arcticus				+/0.2
Luzula parviflora			1/0.7	+/0.2
Lysichitum americanum		8/60.0		
Mahonia nervosa	7/41.5		3.3.6	
Maianthemum dilatatum		+/0.2		
Mycelis muralis	5/15.0	2/1.6	+/0.2	
Osmorhiza chilensis	+/0.2	+/0.2		
Paxistima myrsinites				4/7.5
Peltigera canina				2/1.6
Pinus contorta				5/15.0
Plagiomnium insigne		1/0.7	5/15.0	
Plagiothecium cavifolium		+/0.2		
Polypodium glycyrrhiza			1/0.7	
Polystichum munitum	3/3.6	4/7.5	8/60.0	
Polytrichum piliferum				1/0.7
Pseudotsuga menziesii	9/85.0		5/15.0	
Pteridium aquilinum	+/0.2			
Rhacomitrium canescens				1/0.7
Rhacomitrium heterostichum				1/0.7
Rhytidiadelphus loreus			3/3.6	
Rhytidiadelphus triquetrus	3/3.6			
Rosa gymnocarpa	1/0.7		+/0.2	
Rubus spectabilis		3/3.6	1/0.7	
Rubus ursinus	1/0.7	+/0.2		
Solorina crocea				2/1.6

TABLE 33 Vegetation of test plots used to demonstrate the methods of indicator plant analysis (concluded)

Test plot	Saanich Peninsula	Cowichan Lake	MacMillan Park	Manning Park
Species	**Species cover code/midpoint percent cover**			
Spiraea betulifolia				3/3.6
Stachys cooleyae		2/1.6		
Streptopus amplexifolius			3/3.6	
Symphoricarpos albus	1/0.7			
Tellima grandiflora		+/0.2		
Thuja plicata	3/3.6			
Tiarella laciniata		+/0.2	4/7.5	
Tiarella trifoliata		2/1.6	7/41.5	
Tolmiea menziesii			1/0.7	
Trautvetteria caroliniensis		2/1.6		
Trientalis latifolia	3/3.6			
Trillium ovatum		+/0.2		
Trisetum cernuum			2/1.6	
Tsuga heterophylla			7/41.5	
Vaccinium parvifolium			5/15.0	
Vaccinium scoparium				5/15.0
Total number of species	25	35	27	17
Sum of midpoint percent cover	197.0	160.0	312.7	95.6

TABLE 34 Succession status and soil moisture and soil nutrient regimes of test plots

Attribute	Saanich Peninsula	Cowichan Lake	MacMillan Park	Manning Park
Succession status	mature	early-immature	old growth	late-immature
Soil moisture[1]	moderately dry	wet	moist	very dry
Soil nutrient regime[2]	medium	very rich	very rich	very poor

[1] Inferred from annual water balance and the depth of the growing-season groundwater table.
[2] Inferred from soil chemical properties recommended by Courtin et al. (1987) and Kabzems and Klinka (1987).

Individual-Species Method

The individual-species method is suitable for a fast field diagnosis, particularly in species-poor plant communities with a single dominant (high-cover) species. It is also suitable for diagnosis of microsites (canopy openings, mounds, depressions, disturbed surfaces) occupied usually by several to many individuals of the same species.

In the first step, one or two clearly dominant species or even groups of related species such as lichens or *Sphagnum* spp., with cover greater than 20%, are identified (Figure 6). Then, using Appendix I, desired indicative values of these dominant species are plotted, and the plots are interpreted (Figure 6). For example, in relation to soil moisture, this method suggested a very dry SMR for the Manning Park plot and a relatively narrow range of SMR's for the Saanich Peninsula plot (moderately to slightly dry) and the Cowichan Lake plot (wet to very wet), but failed to provide any information for the MacMillan Park plot because *Achlys triphylla* and *Polystichum munitum* are not indicators of soil moisture (Figure 6). In relation to soil nutrients, it was possible to infer a broad range of SNR's for each test plot (see Table 34).

The results obtained from the individual-species method may or may not be satisfactory depending on the indicative values of a species and the user's objective. If the results are considered unsatisfactory, then another method should be used. While simple and expedient in the field, the individual-species methods often produces incomplete and indefinite results because of incomplete and/or wide-ranging indicative values. However, there is no alternative when dealing with species-poor plant communities. In such situations only the soil examination can help.

Spectral Method

The spectral method of indicator plant analysis requires preparation of a diagnostic summary that utilizes all indicator species for a given attribute. Indifferent species and unclassified species (plants not included in this guide) may be used to determine a reliability ratio (cover of indicator species, divided by cover of all species, and multiplied by 100). The higher this ratio, the more reliable the diagnosis.

Instead of the distribution and cover of individual species, diagnostic summaries show the distribution and frequency of indicator species groups (ISGs) for a site. This requires sorting of species into ISGs for a given attribute. Memberships of species in ISGs recognized in this guide are given in Appendix I. An example of sorting of species of the test plots into soil moisture and soil nutrient ISGs is given in Tables 35 and 36, respectively.

Test plots	Dominant species
Saanich Peninsula	- *Mahonia nervosa* (41.5%)
Cowichan Lake	- *Lysichitum americanum* (60.5%) and *Athyrium filix-femina* (41.5%)
MacMillan Park	- *Polystichum munitum* (60.5%) and *Achlys triphylla* (41.5%)
Manning Park	- lichens (26%)

Dominant species and their midpoint percent cover in the test plots.

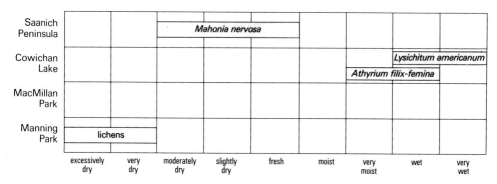

Soil moisture indicator values of the dominant species in the test plots.

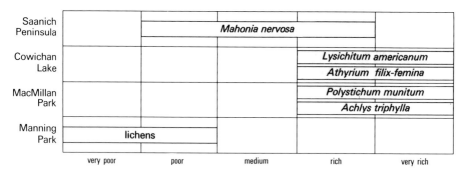

Soil nitrogen indicator values of the dominant species in the test plots.

FIGURE 6   Soil moisture and nutrient diagnosis for test plots using dominant species

TABLE 35   Distribution and midpoint percent cover of soil moisture indicators and ISGs in test plots

Test plot	Saanich Peninsula	Cowichan Lake	MacMillan Park	Manning Park
ISG and species		Midpoint percent cover		
MOIST 1 ISG				
Cladonia apodocarpa				26.5
Cladonia gracilis				7.5
Cladonia pyxidata				0.7
Polytrichum piliferum				0.7
Rhacomitrium canescens				0.7
Rhacomitrium heterostichum				0.7
Number of species				6
Sum of midpoint percent cover				36.8
MOIST 2 ISG				
Arctostaphylos uva-ursi				1.6
Bromus sitchensis		0.2		
Calamagrostis rubescens				7.5
Lonicera ciliosa	0.7			
Lonicera hispidula	0.2			
Peltigera canina				1.6
Rosa gymnocarpa	0.7		0.2	
Spiraea betulifolia				3.6
Number of species	3	1	1	4
Sum of midpoint percent cover	1.6	0.2	0.2	14.3
MOIST 3 ISG				
Adenocaulon bicolor			1.6	
Amelanchier alnifolia				0.7
Corallorhiza maculata	0.2			
Hieracium albiflorum				0.7
Kindbergia oregana	15.0	0.2	7.5	
Lathyrus nevadensis	0.7			
Mahonia nervosa	41.5		3.6	
Paxistima myrsinites				7.5
Rubus ursinus	0.7	0.2		
Trientalis latifolia	3.6			
Number of species	6	2	3	3
Sum of midpoint percent cover	61.7	0.4	12.7	8.9

TABLE 35   Distribution and midpoint percent cover of soil moisture indicators and
ISGs in test plots (continued)

Test plot	Saanich Peninsula	Cowichan Lake	MacMillan Park	Manning Park
ISG and species		Midpoint percent cover		
MOIST 4 ISG				
Acer macrophyllum		0.2		
Adiantum pedatum		0.2		
Blechnum spicant		0.7		
Carex deweyana		0.7		
Carex hendersonii			3.6	
Circaea alpina		0.2		
Claytonia sibirica		0.2		
Disporum hookeri		0.2	3.6	
Dryopteris expansa			1.6	
Equisetum telmateia		0.2		
Galium triflorum	0.7	0.2	26.5	
Luzula parviflora			0.7	0.2
Mycelis muralis	15.0	1.6	0.2	
Osmorhiza chilensis	0.2	0.2		
Rhytidiadelphus loreus			3.6	
Streptopus amplexifolius			3.6	
Tellima grandiflora		0.2		
Tiarella laciniata		0.2	7.5	
Tiarella trifoliata		1.6	41.5	
Tolmiea menziesii			0.7	
Trautvetteria caroliniensis		1.6		
Trillium ovatum		0.2		
Trisetum cernuum			1.6	
Number of species	3	16	12	1
Sum of midpoint percent cover	15.9	8.4	94.7	0.2
MOIST 5 ISG				
Athyrium filix-femina		41.5	0.7	
Conocephalum conicum		0.2		
Cornus sericea	0.7			
Festuca subulata	0.7			
Kindbergia praelonga		3.6		
Leucolepis menziesii		3.6		
Maianthemum dilatatum		0.2		
Plagiomnium insigne		0.7	15.0	
Rubus spectabilis		3.6	0.7	
Stachys cooleyae		1.6		
Number of species	2	8	3	
Sum of midpoint percent cover	1.4	55.0	16.4	

TABLE 35   Distribution and midpoint percent cover of soil moisture indicators and ISGs in test plots (concluded)

Test plot	Saanich Peninsula	Cowichan Lake	MacMillan Park	Manning Park
ISG and species		Midpoint percent cover		
MOIST 6 ISG				
Glyceria elata		0.2		
Lysichitum americanum		60.0		
Number of species		2		
Sum of midpoint percent cover		60.2		
Total number of species	14	29	19	14
Total sum of midpoint percent cover	80.6	124.2	124.0	60.2

TABLE 36   Distribution and midpoint percent cover of soil nitrogen indicators and ISGs in test plots

Test plot	Saanich Peninsula	Cowichan Lake	MacMillan Park	Manning Park
ISG and species		Midpoint percent cover		
NITR 1 ISG				
Arctostaphylos uva-ursi				1.6
Blechnum spicant		0.7		
Cladonia apodocarpa				26.5
Cladonia gracilis				7.5
Cladonia pyxidata				0.7
Corallorhiza maculata	0.2			
Cornus unalaschkensis	3.6			
Gaultheria shallon	15.0			
Hylocomium splendens			7.5	
Paxistima myrsinites				7.5
Polytrichum piliferum				0.7
Rhacomitrium canescens				0.7
Rhacomitrium heterostichum				0.7
Rhytidiadelphus loreus			3.6	
Vaccinium parvifolium			15.0	
Vaccinium scoparium				15.0
Number of species	3	1	3	9
Sum of midpoint percent cover	18.8	0.7	26.1	60.9

TABLE 36   Distribution and midpoint percent cover of soil nitrogen indicators and ISGs in test plots (continued)

Test plot	Saanich Peninsula	Cowichan Lake	MacMillan Park	Manning Park
ISG and species		Midpoint percent cover		
NITR 2 ISG				
Amelanchier alnifolia				0.7
Calamagrostis rubescens				7.5
Circaea alpina		0.2		
Dryopteris expansa			1.6	
Equisetum arvense		0.2		
Hieracium albiflorum				0.7
Lonicera ciliosa	0.7			
Lonicera hispidula	0.2			
Luzula parviflora			0.7	0.2
Mahonia nervosa	41.5		3.6	
Peltigera canina	1.6			
Pteridium aquilinum	0.2			
Rhytidiadelphus triquetrus	3.6			
Rosa gymnocarpa	0.7		0.2	
Rubus ursinus	0.7	0.2		
Spiraea betulifolia				3.6
Trientalis latifolia	3.6			
Number of species	8	3	4	5
Sum of midpoint percent cover	66.2	0.6	6.1	12.7
NITR 3 ISG				
Acer macrophyllum		0.2		
Achlys triphylla	0.7	0.7	41.5	
Adenocaulon bicolor			1.6	
Adiantum pedatum		0.2		
Alnus rubra		26.5		
Athyrium filix-femina		41.5	0.7	
Bromus sitchensis		0.2		
Bromus vulgaris	0.7	0.7		
Carex deweyana		0.7		
Carex hendersonii			3.6	
Claytonia sibirica		0.2		
Conocephalum conicum		0.2		
Cornus sericea	0.7			
Disporum hookeri		0.2	3.6	
Equisetum telmateia		0.2		
Festuca subulata	0.7			
Galium triflorum	0.7	0.2	26.5	
Glyceria elata		0.2		

TABLE 36   Distribution and midpoint percent cover of soil nitrogen indicators and ISGs in test plots (concluded)

Test plot	Saanich Peninsula	Cowichan Lake	MacMillan Park	Manning Park
ISG and species		Midpoint percent cover		
NITR 3 ISG (Concluded)				
Kindbergia praelonga		3.6		
Lathyrus nevadensis	0.7			
Leucolepis menziesii		3.6		
Lupinus arcticus				0.2
Lysichitum americanum		60.0		
Maianthemum dilatatum		0.2		
Mycelis muralis	15.0	1.6	0.2	
Osmorhiza chilensis	0.2	0.2		
Plagiomnium insigne		0.7	15.0	
Polystichum munitum	3.6	7.5	60.0	
Rubus spectabilis		3.6	0.7	
Stachys cooleyae		1.6		
Streptopus amplexifolius			3.6	
Symphoricarpos albus	0.7			
Tellima grandiflora		0.2		
Tiarella laciniata		0.2	7.5	
Tiarella trifoliata		1.6	41.5	
Tolmiea menziesii			0.7	
Trautvetteria caroliniensis		1.6		
Trillium ovatum		0.2		
Trisetum cernuum			1.6	
Number of species	10	29	15	1
Sum of midpoint percent cover	23.7	158.3	208.3	0.2
Total number of species	21	33	22	15
Total sum of midpoint percent cover	95.1	159.6	240.5	73.8

For each plot, percent frequency (Fjk) is calculated for ISGj (eg. fresh and very moist ISG) for a site attribute k (eg. soil moisture), and is calculated according to:

$$Fjk = (\textstyle\sum_{i} Cijk / \sum_{i}\sum_{j} Cijk) \times 100$$

where Cijk is the midpoint percent cover for indicator species i for ISGj for a site attribute k.

The diagnostic (tabular or graphical) summary, in which frequency values less than 0.5% are not given, is referred to as a spectrum because it shows a profile of ISGs for a site. The diagnosis is based on comparison of an unknown spectrum to a series of standard spectra. The objective is to identify the standard spectrum that most closely resembles the unknown spectrum in terms of the distribution and frequencies of ISGs. The premise for the development of standard spectra is that regardless of differences in the life-form and floristic composition, environmentally similar plant communities are expected to have similar spectra.

Using Table 35 as the source of midpoint percent cover, and the formula for calculating percent frequency for ISGs, soil moisture spectra for the test plots are presented in a tabular (Table 37) and graphical format (Figure 7). Tentative standard spectra (Table 38, Figure 8) that are proposed for the soil moisture diagnosis in coastal British Columbia were derived from the analysis of 128 samples of understory vegetation collected by Klinka (1976), Roy (1984), and Kabzems (1985) in mature forest stands. The SMR of these plots was determined from a combination of annual water balances and the depth of the growing-season groundwater table (see the section on Soil Moisture). In this analysis, it was not possible to distinguish between the slightly dry and fresh SMRs. A key has been developed to facilitate identification of actual SMRs (Table 39).

A comparison of the test plot and standard spectra (Figures 7 and 8) and the key to identification of SMR's suggests that the Saanich Peninsula plot has a moderately dry SMR, the Cowichan Lake plot a wet SMR, the MacMillan Park plot a moist SMR, and the Manning Park Plot a very dry SMR. This diagnosis corresponds well with the results of environmental analysis for the test plots (see Table 35).

TABLE 37   Soil moisture spectra for the plots

Test plot	Distribution and frequency (%) of MOIST ISGs						Total cover		Reliability ratio
	1[1]	2	3	4	5	6	Indicators	All species	
Saanich Peninsula		2	76	20	2		81	197	41
Cowichan Lake				7	44	49	124	160	77
MacMillan Park			10	77	13		124	313	40
Manning Park	61	24	15				60	96	62

[1]See Table 12 for explanation of codes (page 22).

It appears that the SNR can be estimated from the frequency of the NITR3 ISG, representing indicators of nitrogen-rich soils. Using Table 36 as the source of midpoint percent cover and the formula for calculating percent frequency for ISGs, the calculated frequencies of the *Tiarella trifoliata* ISGs for the test plots (Table 40) were compared to the standard spectra proposed for the soil nutrient diagnosis in coastal British Columbia (Table 41). The standard spectra were derived from the analysis of 195 samples of understory vegetation compiled and analyzed by Courtin et al. (1987) for a quantitative characterization of SNR's. In this analysis, it was not possible to distinguish between the rich and very rich SNRs.

Very dry

1	2	3	4
21%	36%	42%	1%

Moderately dry

2	3	4	5
6%	88%	5%	1%

Slightly dry and fresh

3	4	5
46%	51%	3%

Moist

2	3	4	5
1%	12%	62%	25%

Very moist

3	4	5	6
5%	52%	41%	2%

Wet

3	4	5	6
2%	36%	29%	33%

FIGURE 7   Soil moisture spectra for test plots. Frequency values for ISGs are shown outside histograms below symbols for ISGs: 1 - MOIST1, 2 - MOIST2, 3 - MOIST3, 4 - MOIST4, 5 - MOIST5, 6 - MOIST6.

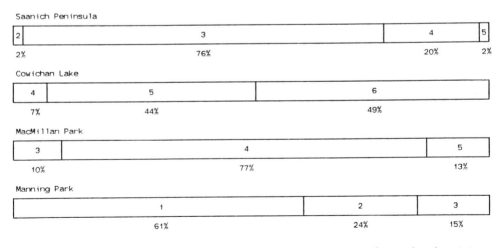

Saanich Peninsula

2	3	4	5
2%	76%	20%	2%

Cowichan Lake

4	5	6
7%	44%	49%

MacMillan Park

3	4	5
10%	77%	13%

Manning Park

1	2	3
61%	24%	15%

FIGURE 8   Tentative standard spectra proposed for diagnosis of actual soil moisture regimes in coastal British Columbia. Frequency values for ISGs are shown outside histograms (see Figure 7).

Comparison of the frequencies of the NITR3 ISG for the test plots to the standard frequency classes suggested that the Saanich Peninsula plot has a medium SNR, the Cowichan Lake and MacMillan Park plots a rich to very rich SNR, and the Manning Park a very poor SNR. This diagnosis agrees with results of environmental analysis for the test plots (see Table 34).

A simplified field version of the spectral method has been described by Green et al. (1984) and Klinka et al. (1984). A computerized version is included in a VTAB program (Emanuel 1986) which performs sorting, frequency calculations, and plotting of spectral histograms. The VTAB is listed under the ID F405 and is available on the UBC General System.

The spectral method is more elaborate and time-consuming than the individual-species method but in return, it provides the best possible ecological summary of a plant community. Spectra have been used by many workers (Raunkiaer 1907; Adamson 1939; Cain 1950; Ellenberg 1950; Prusa 1972; Mueller-Dombois and Ellenberg 1974; Spurr and Barnes 1980) as a simple means for summarizing and interpreting vegetation without presenting individual species. The spectral method effectively exposes trends as well as site heterogeneity (the presence of microsites).

TABLE 38   Tentative standard spectra proposed for diagnosis of soil moisture in coastal British Columbia

Actual SMR	Distribution and frequency (%) of ISGs					
	MOIST1	MOIST2	MOIST3	MOIST4	MOIST5	MOIST6
Number of plots	7	21	31	23	27	19
Very dry	21[2]	36	42	1		
	(24)	(24)	(21)	(2)		
Moderately dry		6	88	5	1	
		(8)	(11)	(5)	(1)	
Slightly dry and fresh			46	51	3	
			(28)	(27)	(4)	
Moist		1	12	62	26	
		(2)	(12)	(18)	(17)	
Very moist			5	52	42	2
			(10)	(18)	(19)	(3)
Wet			2	36	29	33
			(4)	(27)	(22)	(22)

[1] See Table 12 for explanation of codes. (page 22)
[2] Means; numbers in parenthesis are standard deviations.

TABLE 39 Tentative key to actual soil moisture regimes in coastal British Columbia

1a	Spectrum that has the MOIST1 ISG present and/or the MOIST2 ISG well represented	**very dry**
1b	Other spectra	2
2a	Spectrum that has the MOIST6 ISG well represented or dominant	**wet**
2b	Other spectra	3
3a	Spectrum that has the MOIST3 ISG dominant or well represented	4
3b	Other spectra	5
4a	Spectrum that may have the MOIST2, MOIST4, and MOIST5 ISGs present but poorly represented	**moderately dry**
4b	Spectrum that has the MOIST4 ISG dominant or well represented, and if present, other ISGs poorly represented	**slightly dry or fresh**
5a	Spectrum that has the MOIST4 ISG dominant, the MOIST5 ISG well represented, and if present, other ISGs poorly represented	**moist**
5b	Spectrum that has the MOIST5 ISG dominant or well represented and the MOIST6 ISG is usually present but poorly represented	**very moist**

Explanation to the key:

- a poorly represented ISG has frequency <15%
- a well represented ISG has frequency >15%
- a dominant ISG has the highest frequency in a particular spectrum

TABLE 40 Soil nutrient spectra for test plots

Test plot	Distribution and frequency (%) of ISGs			Total cover		Reliability ratio
	NITR1[1]	NITR2	NITR3	Indicators	All species	
Saanich Peninsula	20	55	25	95	197	48
Cowichan Lake	1	-	99	160	160	100
MacMillan Park	11	2	87	289	313	77
Manning Park	83	17	-	75	96	77

[1] See Table 20 for explanation of codes. (page 30)

TABLE 41   Tentative standard frequency classes of NITR3 ISG proposed for diagnosis of actual soil nutrient regimes in coastal British Columbia

Actual SNR	Frequency (%) of NITR3 ISG
Very poor	<6
Poor	6-18
Medium	19-42
Rich and very rich	>42

Index Method

The index method of indicator plant analysis presented here basically involves weighting a species (Minore 1972; Minore and Carkin 1974, 1978) or the number or cover of a species (Rowe 1956) by a numerical value, which is assigned to each species and connotes a class of an environmental gradient. The numerical value can be simply a sequential number of a linear scale (1, 2, 3, 4, ...) - so-called 'ecological indicator species value' (Ellenberg 1974) - or a weight value of a selected exponential scale. Weighted averages have been used in interpreting vegetation-environment relationships and in both direct and indirect gradient analyses as an aid in correlating the pattern in ordination with that of environmental factors (Whittaker 1954, 1978; Curtis 1959; Ellenberg 1974; Landolt 1977; Persson 1981; Kenkel 1987).

For each plot, an **indicator index** (IIk) is calculated for a site attribute k according to:

$$IIk = (\Sigma_i Cij \times Zj / \Sigma_i Cik) \times 10$$

where Cij is the midpoint percent cover for indicator species i for ISGj (eg., excessively dry to very dry ISG), Zj is the weight value for ISGj, and Cik is the midpoint percent cover for indicator species i for site attribute k (e.g., soil moisture).

In this guide, for soil moisture the weight values followed the 2^{j-1} scale, for soil nutrients the 3^{j-1} scale to increase contrast. Indifferent species and species with unknown indicative values are excluded from calculations, but they may be used to determine the reliability ratio.

Soil moisture and soil nutrient diagnosis is based on comparison of the calculated II for a site against the classes of standard IIs. The objective is to identify the class that includes the calculated II. Using Tables 35 and 36 as the source of midpoint percent cover, the calculation of soil moisture and soil nutrient IIs for the test plots is shown in Tables 42 and 43, respectively. Standard IIs for the diagnosis of SMRs and SNRS (Table 44) were derived from the analysis of the same data used to develop tentative standard spectra. These standard IIs are tentative and require further testing. As it was explained earlier, it was not possible to distinguish between the slightly dry and fresh SMRs and rich and very rich SNRs.

TABLE 42   Calculation of soil moisture indicator index for test plots

Indicator species group		Test plot			
		Saanich Peninsula	Cowichan Lake	MacMillan Park	Manning Park
Sum of the midpoint percent cover of species in indicator species groups					
MOIST1[1]		-	-	-	36.8
MOIST2		1.8	0.2	0.2	14.3
MOIST3		61.7	0.4	12.7	8.9
MOIST4		15.9	8.4	94.7	0.2
MOIST5		1.4	55.0	16.4	-
MOIST6		-	60.2	-	-
Total (A)		80.6	124.2	124.0	60.2
Sum of the midpoint percent cover x weight value					
MOIST1	1 x	-	-	-	36.8
MOIST2	2 x	3.6	0.4	0.4	28.6
MOIST3	4 x	246.8	1.6	50.8	35.6
MOIST4	8 x	127.2	67.2	757.6	1.6
MOIST5	16 x	22.4	880.0	262.4	-
MOIST6	32 x	-	1926.4	-	-
Total (B)		400.0	2875.6	1071.2	102.6
Soil moisture indicator index [(B : A) x 10]		**50**	**232**	**86**	**17**

[1] See Table 12 for explanation of codes (page 22).

Comparison of test plot and standard IIs suggested that the Saanich Peninsula plot has a moderately dry SMR, the Cowichan Lake plot a wet SMR, the MacMillan Park plot a moist SMR, and the Manning Park plot a very dry SMR. Comparison of soil nutrient IIs suggested that the Saanich Peninsula plot has a medium SNR, the Cowichan Lake and MacMillan Park plots a rich to very rich SNR, and the Manning Park plot a very poor SNR. This diagnosis agrees with results of environmental analysis for the test plots (see Table 34).

Like the spectral method, the index method is time-consuming, but it does provide a weighted average, expressing the intensity of an environmental factor for a site. A VTAB program (Emanuel 1986) includes an index command that calculates IIs. The index method works best when applied to several mutually exclusive classes of a single environmental factor, such as temperature-based or precipitation-based climatic regimes, soil moisture regime, or soil nutrient regimes. Compared to the spectral method, which shows the distribution and frequency of several ISGs, the indicator index is not able to express site heterogeneity (the presence of microsites) - it merely suggests the position of a site on the environmental gradient.

TABLE 43   Calculation of soil nutrient indicator index for test plots

		Test plot		
Indicator species group		Saanich Peninsula	Cowichan Lake	MacMillan Park	Manning Park
Sum of the midpoint percent cover of species in indicator species groups					
NITR1[1]		18.8	0.7	26.1	60.9
NITR2		52.6	0.6	6.1	12.7
NITR3		23.7	158.3	208.3	0.2
Total (A)		95.1	159.6	240.5	73.8
Sum of the midpoint percent cover x weight value					
NITR1	1 x	18.8	0.7	26.1	60.9
NITR2	3 x	157.8	1.8	18.3	38.1
NITR3	9 x	213.3	1424.7	1874.7	1.8
Total (B)		389.9	1427.2	1919.1	100.8
Soil nutrient indicator index [(B : A) x 10]		**41**	**89**	**80**	**14**

[1] See Table 20 for explanation of codes. (page 30)

TABLE 44   Tentative standard indicator index classes proposed for diagnosis of actual
soil moisture and soil nutrient regimes in coastal British Columbia

Actual SMR	Indicator index	Actual SNR	Indicator index
Very dry	<35	Very poor	<22
Moderately dry	35- 54	Poor	22-31
Slightly dry and fresh	55- 75	Medium	32-48
Moist	76-102	Rich and very rich	>48
Very moist	103-145		
Wet	>145		

6

DISTRIBUTION AND ECOLOGICAL CHARACTERISTICS OF INDICATOR PLANTS

This chapter presents a brief commentary and colour illustration for each of the selected indicator plants. The species are arranged in alphabetical order according to their Latin names. Some ecologically similar species within general have been treated together. The commentary for each species includes an outline of salient distributional and ecological characteristics. The reader is reminded that some of these characteristics may not be valid outside coastal British Columbia.

The geographic elements of Daubenmire (1978) as modified by Krajina et al. (1982) were used to describe the geographic distribution of each species. These geographic elements were determined by plotting on maps the ranges of many species. This procedure exposes distribution patterns that may be generalized into a limited number of geographical areas (elements) (Table 45, Figure 9). The geographic elements are given self-explanatory names, and their use allows for a more efficient description of plant distribution than would the use of many local geographic names.

Life-form describes the general growth habit of a plant species by using a limited number of well defined and easily recognizable morphological features that appear to be correlated with climatic and/or soil factors. The conventional approach to the classification of life forms is based upon the vertical position of over-wintering buds, leaf form, leaf persistence, and/or anatomical structure. For practical reasons, a purely physiognomic approach, based on appearance, is used here (Table 46). This approach simplifies informal plant identification and assists in the characterization and interpretation of vegetation. The nomenclature is self-explanatory; the symbols are used in Appendix II.

Plant-light relationships were evaluated according to shade tolerance, used here as the ability of a plant species to survive in light of low intensity (after Shirley 1943; Baker 1949). This ability, however, may differ in different temperature-moisture environments (Krajina 1965; Krajina et al. 1982; Minore 1987) and different developmental stages (Krajina 1965; Krajina et al. 1982). In this guide, five relative qualitative classes were used: very shade-tolerant (usually shade-requiring species), shade-tolerant, shade-tolerant/intolerant, shade-intolerant, and very shade-intolerant (usually a full light-requiring species). Consequently, there is no intent to imply that the shade tolerance of any two species is comparable in an absolute sense.

TABLE 45   Geographic elements used to describe distribution of plants

Geographic element
North American[1]
Pacific
Cordilleran
Central
Atlantic
transcontinental (Pacific, Cordilleran, Central, and Atlantic)
Asian
European
South American
circumpolar
cosmopolitan

[1]   The Pacific and Cordilleran regions combined are referred to as Western North America; the Central region is referred to as Central North America; the Central and Atlantic regions combined are referred to as Eastern North America. Thus, the geographic distribution of species is expressed by a combination of appropriate adjectives, e.g., a Pacific North American forb, transcontinental North American moss, Asian and Western North American fern distributed more in the Pacific than Cordilleran region, circumpolar sedge transcontinental in North America.

TABLE 46   Life-forms used to describe appearance of plants

Symbol	Life-form
CNTR	coniferous trees
BLTR	broad-leaved trees
EGSH	evergreen shrubs
DCSH	deciduous shrubs
FERN	ferns and allies (horsetails, clubmosses, and 'selaginellas')
GRAM	graminoids (grasses, sedges, and rushes)
FORB	forbs (non-graminoid herbs)
PASA	parasites and saprophytes
MOSS	mosses
LVRT	liverworts
LCHN	lichens

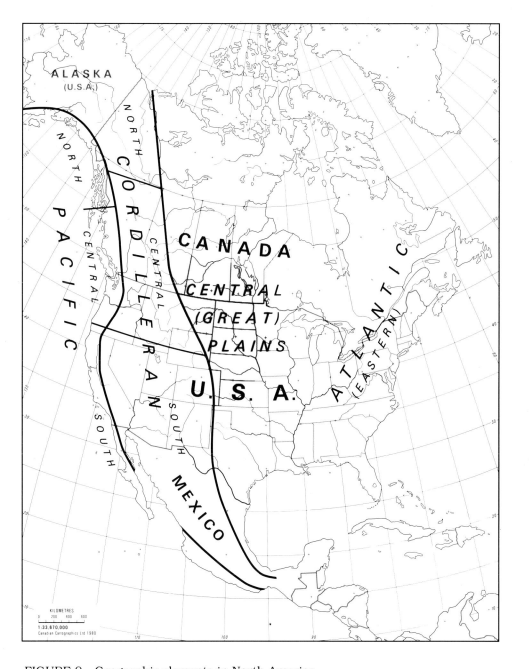

FIGURE 9   Geographic elements in North America

On pages 66 to 248 inclusive, accepted scientific names are printed in bold italics and synonyms in light italics. If more than one species are treated in one block, the illustrated species is listed first.

Abies amabilis
Pacific silver fir, amabilis fir

Pinaceae
(Pine family)

A very shade-tolerant, submontane to subalpine, Western North American evergreen conifer distributed more in the Pacific than the Cordilleran region (absent on Queen Charlotte Islands). Occurs in maritime to submaritime sub-alpine boreal and summer-wet cool mesothermal climates on fresh to very moist soils; its occurrence increases with increasing elevation and precipitation, and decreases with increasing latitude and continentality. Grows in pure or mixed-species stands (usually with western or mountain hemlock) on water-shedding and water-receiving sites. Regenerates underneath closed-canopy stands, particularly on mycorrhizal Mors. (The mycorrhizae may explain this species' tolerance of nutrient-poor sites.) Most productive on submontane, fresh to moist, nutrient-rich (seepage) sites within wet cool mesothermal climates. Characteristic of wet maritime forests.

Abies grandis
Grand fir

Pinaceae
(Pine family)

A shade-tolerant to shade-intolerant, submontane to montane, Western North American evergreen conifer distributed less in the Pacific than the Cordilleran region. Occurs in cool temperate and cool mesothermal climates; its occurrence decreases with increasing latitude, precipitation, and elevation. Grows in mixed-species stands (usually with Douglas-fir or western redcedar) on water-shedding and water-receiving sites. Tolerates fluctuating groundwater tables. Most productive on submontane, fresh to moist, nutrient-rich (alluvial and seepage) sites. Characteristic of nutrient-rich sites.

Abies lasiocarpa
Subalpine fir

Pinaceae
(Pine family)

A shade-tolerant to shade-intolerant, montane to subalpine, Western North American evergreen conifer distributed less in the Pacific than the Cordilleran region. Occurs predominantly in continental boreal climates; its occurrence increases with both elevation and continentality. Grows in mixed-species stands (usually with Engelmann spruce, Pacific silver fir, or mountain hemlock) on leeward slopes of Vancouver Island and Coastal Mountain Ranges. There are occasional pure stands at high elevations on exposed outcrops of base-rich rocks, or on valley bottoms affected by cold air drainage. On sites where it is shade-tolerant, it regenerates under closed-canopy stands, particularly on mycorrhizal Mors. (The mycorrhizae here may explain its tolerance of nutrient-poor soils.) Most productive on montane, fresh to moist, nutrient-rich (seepage) sites within wet cool temperate climates. Characteristic of continental boreal forests.

Acer circinatum
Vine maple

Aceraceae
(Maple family)

A shade-tolerant to shade-intolerant, submontane to montane, Pacific North American deciduous shrub (rare on Vancouver Island). Occurs in maritime to submaritime cool mesothermal climates on fresh to very moist, nitrogen-rich soils; its occurrence decreases with increasing elevation and continentality. Plentiful and persistent in open-canopy forests and clearings on water-receiving (alluvial, seepage, and stream-edge) sites; dominant in primary successional stages on water-shedding sites with fragmental colluvial soils.

Regenerates vigorously from stump sprouts; it hinders natural regeneration and growth of shade-intolerant conifers. Frequently grows with *Polystichum munitum*. Characteristic of Moder and Mull humus forms.

Acer glabrum
Acer douglasii
Douglas maple

Aceraceae
(Maple family)

A shade-tolerant to shade-in-tolerant, montane to subalpine, Western North American deciduous shrub distributed more in the Cordilleran than the Pacific region. Occurs on nitrogen-rich, water-shedding and water-receiving sites within continental boreal, cool temperate, cool semiarid, and occasionally cool mesothermal climates. Scattered throughout coastal British Columbia; its occurrence increases with increasing continentality. Common on eastern Vancouver Island and in Skeena River valley. Grows with vine maple in the southern coast-interior ecotone. Like *A. circinatum*, it inhabits open-canopy forests, clearings, and primary succession stages on fragmental colluvial soils. Regenerates abundantly from stump sprouts; it hinders regeneration and growth of shade-intolerant conifers. Characteristic of Moder and Mull humus forms.

Acer macrophyllum
Broad-leaved maple, bigleaf maple

Aceraceae
(Maple family)

A shade-intolerant, submontane to montane, Western North American deciduous broad-leaved tree distributed more in the Pacific than the Cordilleran region. Occurs in maritime to submaritime cool mesothermal climates on fresh to very moist, nitrogen-rich soils (Moder and Mull humus forms). Its occurrence decreases with increasing elevation, latitude, and continentality. Common in pure or mixed-species stands (usually with red alder or black cottonwood) on alluvial, seepage, and stream-edge sites; occasional on water-shedding sites; dominant in primary succession on fragmental colluvial soils. This fast-growing tree regenerates abundantly from stump sprouts in clearings, thus hindering regeneration and growth of conifers. Its calcium-rich bark supports well developed corticolous moss communities. Characteristic of young-seral forests.

Achillea lanulosa
Achillea millefolium
Western yarrow

Asteraceae
(Aster family)

A very shade-intolerant, submontane to montane, transcontinental North American forb. Occurs on very dry to moderately dry, nitrogen-medium soils; its occurrence decreases with increasing elevation and precipitation. Inhabits exposed mineral soils on water-shedding sites (often with melanized soils) within boreal, temperate, cool semiarid, and mesothermal climates; most frequent in early stages of secondary succession. Characteristic of disturbed sites.

Achlys triphylla
Vanillaleaf

Berberidaceae
(Barberry family)

A shade-tolerant, submontane to montane Western North American forb distributed more in the Pacific than the Cordilleran region. Occurs in maritime to submaritime cool mesothermal climates on nitrogen-rich soils. Its occurrence decreases with increasing latitude, elevation, and continentality; plentiful on Vancouver Island, sparse on coastal mainland. Most frequent on water-shedding and water-receiving sites; commonly associated with *Polystichum munitum*. Characteristic of Moder and Mull humus forms.

Actaea rubra
Baneberry

Ranunculaceae
(Buttercup family)

A shade-tolerant, submontane to subalpine, transcontinental North American forb. Occurs on fresh to very moist, nitrogen-rich soils within boreal, cool temperate, and cool mesothermal climates. Its occurrence increases with increasing precipitation and continentality. Occasional in broad-leaved forests on water-receiving (alluvial, floodplain, seepage, and stream-edge) sites. A nitrophytic species characteristic of Moder and Mull humus forms.

Adenocaulon bicolor
Pathfinder

Asteraceae
(Aster family)

A shade-tolerant/intolerant, submontane to montane, North American forb distributed in the Pacific (most frequently), Cordilleran, and Central regions. Occurs in cool temperate and cool mesothermal climates on moderately dry to fresh, nitrogen-rich soils; its occurrence decreases with increasing latitude and elevation. Sporadic in coniferous forests, scattered in broad-leaved forests on water-shedding sites. Often associated with *Polystichum munitum* and *Tiarella trifoliata*. A nitrophytic species characteristic of Moder and Mull humus forms.

Adiantum pedatum
Maidenhair fern

Adiantaceae
(Maidenhair fern family)

A shade-tolerant/intolerant, submontane to montane, East Asian and transcontinental North American fern. Occurs in cool temperate and cool mesothermal climates on fresh to very moist, calcium-rich and nitrogen-rich soils; its occurrence decreases with increasing elevation. Scattered, occasionally plentiful, on water-receiving (alluvial, floodplain, seepage, and stream-edge) sites. Often grows with *Oplopanax horridus*, *Polystichum munitum*, and *Tiarella trifoliata*. Characteristic of Moder and Mull humus forms.

Agropyron spicatum
Bluebunch wheatgrass

Poaceae
(Grass family)

A shade-intolerant, montane, Cordilleran North American grass (rare in the Pacific region). Occurs in cool temperate and cool semiarid climates on excessively dry to very dry, circumneutral to weakly alkaline, nitrogen-rich soils. Sporadic in the grassy understory of open-canopy ponderosa pine or Douglas-fir forests, dominant in grassland communities on shallow, water-shedding sites in the south-eastern part of the coast-interior ecotone. A nitrophytic species characteristic of Moder and Mull humus forms.

Agrostis aequivalvis
Podagrostis aequivalvis
Northern bentgrass

Poaceae
(Grass family)

A very shade-intolerant, submontane to montane, Western North American grass distributed more in the Pacific than the Cordilleran region. Occurs in cool mesothermal climates on very moist to wet nitrogen-medium soils. Its occurrence decreases with increasing elevation and continentality. Grows in non-forested and open-canopy semi-terrestrial communities on water-receiving and water-collecting sites. Characteristic of wetlands.

Aira caryophyllea
Silver hairgrass
Aira praecox
Early hairgrass

Poaceae
(Grass family)

Very shade-intolerant, submontane to montane, European grasses introduced to Pacific and Atlantic North America. Both species occur in maritime to submaritime summer-dry cool mesothermal climates on very dry to moderately dry soils, and, *A. praecox*, on nitrogen-poor soils. Scattered in the open and open-canopy coniferous forests on very shallow and strongly drained soils of rock outcrops. Characteristic of moisture-deficient sites.

Alectoria vancouverensis

Usneaceae

A shade-intolerant, submontane to montane, Pacific North American lichen. Occurs in summer-dry cool mesothermal climates; its occurrence decreases with increasing elevation and continentality. Inhabits branches and stems of conifers. Characteristic of exposed, open-canopy, mature forests on upland sites.

Allium acuminatum
Hooker's onion

Liliaceae
(Lily family)

A very shade-intolerant, montane, Western North American forb distributed equally in the Pacific and Cordilleran regions. Occurs in cool temperate and cool mesothermal climates very dry to moderately dry and nitrogen-medium soils. Its occurrence decreases with increasing latitude, precipitation, and continentality. Scattered in the open and in open-canopy forests on water-shedding sites with shallow soils. Characteristic of moisture-deficient sites.

Allium cernuum
Nodding onion

Liliaceae
(Lily family)

A shade-intolerant, montane to subalpine, transcontinental North American forb. Occurs on nitrogen-medium, water-shedding sites within boreal, temperate, and cool mesothermal climates. Its occurrence decreases with increasing precipitation and latitude. Occasional in early-seral communities on shallow soils of rock outcrops; frequently inhabits exposed mineral soils. Characteristic of summer-dry temperate and mesothermal forests.

Allotropa virgata
Candystick

Monotropaceae
(Indian Pipe family)

A shade-intolerant, submontane to montane, Western North American parasite distributed more in the Pacific than the Cordilleran region. Occurs in cool temperate and cool mesothermal climates on very dry to moderately dry, nitrogen-poor soils (Mor humus forms). Its occurrence decreases with increasing latitude, precipitation, and continentality. Sporadic in open-canopy Douglas-fir forests on water-shedding sites. Commonly associated with *Hylocomium splendens* and *Pleurozium schreberi*. An oxylophytic species characteristic of moisture-deficient sites.

Alnus rubra
Alnus oregona
Red alder

Betulaceae
(Birch family)

A shade-intolerant, submontane to montane, Pacific North American deciduous broad-leaved tree. An abundant species that grows in cool mesothermal climates on nitrogen-rich soils (Moder and Mull humus forms); its occurrence decreases with increasing elevation and continentality. Forms dense stands in the initial stages of primary succession on floodplains or secondary succession on water-shedding sites. Persists along streams and on water-collecting sites, usually associated with *Lysichitum americanum*; tolerates fluctuating groundwater tables. This fast-growing tree regenerates abundantly from seed on exposed mineral soil, and from stump sprouts following cutting. May hinder regeneration and growth of conifers. Symbiosis with nitrogen-fixing Actinomycetes enhances the supply of available soil nitrogen. Suitable as a temporary nurse species for shade-tolerant conifers, especially on nitrogen-deficient sites; however, it may decrease both soil pH and base content of some soils. Characteristic of young-seral mesothermal forests.

Alnus sinuata
Alnus viridis ssp. *sinuata*
Sitka alder

Betulaceae
(Birch family)

A shade-tolerant, submontane to alpine, Asian and North American deciduous shrub distributed equally in the Pacific and Cordilleran regions. Occurs on fresh to very moist, nitrogen-rich soils (Moder and Mull humus forms) within boreal, cool temperate, and cool mesothermal climates; its occurrence increases with increasing elevation. Most common in open-canopy forests on water-receiving (alluvial, floodplain, seepage and stream-edge) sites; dominates high-elevation, steep-gradient, stream-edge, and avalanche-track communities. Regenerates abundantly on exposed mineral soil or from stump sprouts; tolerates fluctuating groundwater tables. Symbiosis with nitrogen-fixing Actinomycetes enhances the supply of available soil nitrogen. A very suitable temporary or long-term nurse species for coniferous crops (Engelmann spruce, lodgepole pine, and white spruce) on nitrogen-deficient sites. Characteristic of disturbed sites.

Amelanchier alnifolia
Saskatoon

Rosaceae
(Rose family)

A shade-tolerant to shade-intolerant, submontane to montane, North American deciduous shrub distributed equally in the Pacific, Cordilleran, and Central regions. Occurs on moderately dry to fresh, nitrogen-medium soils within boreal, cool temperate, cool semiarid, and cool mesothermal climates. Its occurrence increases with increasing continentality, and decreases with increasing precipitation and elevation. Common to scattered in clearings and open-canopy Douglas-fir and lodgepole pine stands on water-shedding sites. Characteristic of young-seral forests on disturbed sites.

Anaphalis margaritacea
Pearly-everlasting

Asteraceae
(Aster family)

A shade-intolerant, submontane to subalpine, Asian and transcontinental North America forb. Occurs on water-shedding sites within alpine tundra, boreal, cool temperate, and cool mesothermal climates. Inhabits exposed mineral soil on cutover sites, clearings, and waysides. Under these circumstances often dominates initial stages of secondary succession. Characteristic of disturbed sites.

Andromeda polifolia
Bog-rosemary

Ericaceae
(Heath family)

A very shade-intolerant, sub-montane to montane, circum-polar forb (transcontinental in North America). Occurs on wet to very wet, nitrogen-poor soils (Mor humus forms) within boreal, cool temperate and cool mesothermal climates. Inhabits water-collecting sites with large accumulations of peat. Occasional amidst *Sphagnum* species in peat bogs. An oxylo-phytic species characteristic of nutrient-poor wetlands.

Angelica genuflexa
Kneeling angelica

Apiaceae
(Parsley family)

A shade-tolerant/intolerant, submontane to montane, Asian and Western North American forb distributed more in the Pacific than the Cordilleran region. Occurs in cool temperate and cool mesothermal climates on wet to very wet, nitrogen-rich soils (Moder and Mull humus forms); its occurrence decreases with increasing elevation and continentality. Common in open-canopy red alder, Sitka spruce, and western redcedar stands on water-collecting sites with gleysolic or organic soils. Often associated with *Athyrium filix-femina* and *Lysichitum americanum*. Inhabits depressions with a surface groundwater table. A nitro-phytic species characteristic of nutrient-rich wetlands.

Antennaria neglecta
Field pussytoes

Asteraceae
(Aster family)

A shade-tolerant/intolerant, montane to subalpine, transcontinental North American forb. Occurs in continental boreal and cool temperate climates on very dry to moderately dry soils; its occurrence increases with increasing continentality. Sporadic in non-forested communities or open-canopy forests on water-shedding sites in the coast interior ecotone. Occasionally inhabits exposed mineral soil. Characteristic of moisture-deficient sites.

Apocynum androsaemifolium
Spreading dogbane

Apocynaceae
(Dogbane family)

A shade-intolerant, submontane to montane, transcontinental North American forb distributed more in the Cordilleran than the Pacific region. Occurs on very dry to moderately dry, nitrogen-medium soils within boreal, temperate, and cool mesothermal climates. Its occurrence increases with increasing continentality and temperature, and decreases with increasing elevation and precipitation. Usually inhabits unshaded and strongly drained, water-shedding sites with exposed mineral soil or thin Mor humus forms. Characteristic of moisture-deficient sites.

Aquilegia formosa
Red columbine

Ranunculaceae
(Buttercup family)

A shade-tolerant/intolerant, montane to subalpine, Western North American forb distributed equally in the Pacific and Cordilleran regions. Occurs on fresh to very moist, nitrogen-rich soils within subalpine boreal, temperate, and cool mesothermal climates; its occurrence decreases with increasing elevation. Scattered in broad-leaved forests on flooded sites, often inhabits exposed mineral soils in early-seral communities on water-receiving sites. A nitrophytic species characteristic of Moder and Mull humus forms.

Aralia nudicaulis
Wild sarsaparilla

Araliaceae
(Ginseng family)

A shade-tolerant/intolerant, montane, transcontinental North American forb (absent from the Pacific region). Occurs in continental montane boreal and cool temperate climates on fresh to very moist, nitrogen-rich soils. Scattered to plentiful in continental forests (persists in clearings); sporadic on water-shedding and water-receiving sites in the coast-interior ecotone. Often associated with *Disporum hookeri, Gymnocarpium dryopteris, Oplopanax horridus, Smilacina racemosa,* and *S. stellata.* A nitrophytic species characteristic of Moder and Mull humus forms.

Arbutus menziesii
Pacific madrone, arbutus

Ericaceae
(Heath family)

A shade-intolerant, submontane to montane, Western North American evergreen broad-leaved tree distributed more in the Pacific than the Cordilleran region. Occurs in maritime summer-dry cool mesothermal climates on very dry to moderately dry soils. Restricted to water-shedding sites on southeastern Vancouver Island, Gulf Islands, and adjacent coastal mainland; its occurrence decreases with increasing latitude, elevation, and continentality. Occasional in pure or mixed-species young-seral stands (usually with Garry oak or Douglas-fir) on strongly drained sites. Commonly associated with *Gaultheria shallon*. Characteristic of moisture-deficient sites.

Arctostaphylos columbiana
Hairy manzanita

Ericaceae
(Heath family)

A shade-intolerant, submontane to montane, Pacific North American evergreen shrub. Occurs in maritime summer-dry cool mesothermal climates on very dry to moderately dry, nitrogen-poor soils (Mor humus forms). Occasional in open-canopy, young-seral Douglas-fir forests, more frequent in the open and in clearings, on shallow, strongly drained soils on rock outcrops and upper slopes. An oxylophytic species characteristic of moisture-deficient sites.

Arctostaphylos uva-ursi
Kinnikinnick

Ericaceae
(Heath family)

A shade-tolerant/intolerant, submontane to subalpine, circumpolar evergreen shrub (transcontinental in North America). Occurs on very dry to moderately dry, nitrogen-poor soils (Mor humus forms) within boreal, temperate, and cool mesothermal climates; its occurrence increases with increasing continentality. Common in open-canopy, young-seral lodgepole pine forests on shallow soils, soils on rock outcrops and strongly drained coarse-skeletal soils on water-shedding sites. Often associated with *Gaultheria shallon, Pleurozium schreberi*, and lichens. Characteristic of moisture-deficient sites.

Arnica cordifolia
Heart-leaved arnica

Asteraceae
(Aster family)

A shade-tolerant/intolerant, montane to subalpine, Western North American forb distributed mainly in the Cordilleran region, marginally in the Pacific and Central regions. Occurs in continental boreal and cool temperate climates on moderately dry to fresh, nitrogen-medium soils; its occurrence increases with increasing elevation and continentality. Grows in meadow-like communities and in open-canopy coniferous forests on high-elevation water-shedding sites in the coast-interior ecotone. Often inhabits exposed mineral soils. Characteristic of continental forests.

Arnica latifolia
Mountain arnica

Asteraceae
(Aster family)

A shade-intolerant, montane to subalpine, Asian and Western North American forb distributed equally in the Pacific and Cordilleran regions. Occurs in alpine tundra and boreal climates on fresh to very moist, nitrogen-medium soils. Plentiful to abundant in meadow-like communities, and scattered in open-canopy coniferous forests on high-elevation water-shedding and water-receiving (seepage and stream-edge) sites. Often inhabits exposed mineral soil. Characteristic of alpine and subalpine communities.

Aruncus dioicus
Aruncus sylvester
Goatsbeard

Rosaceae
(Rose family)

A shade-tolerant/intolerant, submontane to subalpine, circumpolar forb distributed in the Pacific, Cordilleran (less frequently), and Atlantic North America. Occurs on fresh to very moist, nitrogen-rich soils within boreal, cool temperate, and cool mesothermal climates. Commonly inhabits exposed mineral soil on water-receiving flooded sites. Scattered in early-seral herbaceous communities in the proximity of intermittent or permanent waterways and run-off channels on steep, often shallow and rocky, colluvial slopes; less frequent in open-canopy, steep-gradient, stream-edge forests. A nitrophytic species characteristic of Moder and Mull humus forms.

Asarum caudatum
Wild ginger

Aristolochiaceae
(Birthwort family)

A shade-tolerant, submontane to montane, Western North American forb distributed in the Pacific and Cordilleran regions. Occurs in cool temperate and cool mesothermal climates on fresh and moist, nitrogen-rich soils. Its occurrence increases with increasing continentality, and decreases with increasing latitude and elevation. Sparse in coniferous forests, scattered in broad-leaved forests on water-receiving (alluvial, colluvial, floodplain, and seepage) sites. Often associated with *Athyrium filix-femina*, *Oplopanax horridus*, and *Rubus parviflorus*. A nitrophytic species characteristic of Moder and Mull humus forms.

Asplenium trichomanes
Maidenhair spleenwort

Aspleniaceae
(Spleenwort family)

A shade-tolerant/intolerant, montane to subalpine, circumpolar fern (transcontinental in North America). Occurs on water-shedding and water-receiving sites within subalpine boreal, cool temperate, and cool mesothermal climates. Sporadic in shrub or forest communities on steep colluvial slopes where it inhabits shaded, humid microsites on rocks, boulders, and stones that have very shallow, friable soils high in organic matter. Characteristic of colluvial sites.

Aster ciliolatus
Fringed aster

Asteraceae
(Aster family)

A shade-tolerant/intolerant, montane trans-continental North American forb (absent in the Pacific region). Occurs in continental montane boreal and cool temperate climates on moderately dry to fresh, nitrogen-medium soils. A common interior species; scattered in young-seral forests on water-shedding sites in the eastern coast-interior ecotone. Often associated with *Aster conspicuus, Calamagrostis rubescens*, and *Spiraea betulifolia.* Characteristic of continental forests.

Aster conspicuus
Showy aster

Asteraceae
(Aster family)

A shade-tolerant/intolerant, montane to subalpine, North American forb distributed more in the Cordilleran than the Central region (rare in the Pacific region). Occurs in continental boreal and cool temperate climates on moderately dry to fresh, circumneutral to weakly alkaline, nitrogen-rich soils. A common interior species on water-shedding sites; scattered to plentiful in Douglas-fir, lodgepole pine, and ponderosa pine stands in the eastern coast-interior ecotone. Often associated with *Aster ciliolatus, Calamagrostis rubescens*, and *Spiraea betulifolia.* A nitrophytic species characteristic of Moder and Mull humus forms.

Athyrium filix-femina
Lady fern

Aspleniaceae
(Spleenwort family)

A shade-tolerant, submontane to montane, circumpolar fern distributed in Pacific, Cordilleran, and Atlantic North America. Occurs on very moist to wet, nitrogen-rich soils within boreal, temperate, and mesothermal climates. Plentiful to abundant (occasionally dominant) in non-forested communities or forest understories on water-receiving (alluvial, floodplain, seepage, and stream-edge) sites; often inhabits water-collecting (swamps and fens) sites. A nitrophytic species characteristic of Moder and Mull humus forms.

Atrichum selwynii
Atrichum undulatum

Polytrichaceae

Shade-tolerant, submontane to montane, Western North American mosses distributed in the Pacific, Cordilleran, and Central (marginally) regions. Both species occur in cool temperate and cool mesothermal climates on fresh to very moist, nitrogen-rich soils. Their occurrence decreases with increasing elevation and continentality. Occasional in closed-canopy forests on water-receiving sites. Characteristic of microsites with exposed, friable, mineral soils.

Aulacomnium palustre

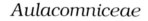
Aulacomniceae

A shade-intolerant, submontane to subalpine, cosmopolitan moss (transcontinental in North America). Occurs on wet to very wet, nitrogen-medium soils within boreal, temperate, and cool mesothermal climates. Scattered in non-forested semiterrestrial communities on water-receiving sites with gleysolic or organic soils. Occasional in depressions in open-canopy forests where the growing-season groundwater table is at or above the ground surface except in the driest months of summer. Characteristic of wetlands.

Barbilophozia floerkei
Barbilophozia lycopodioides

Lophoziaceae

Shade-tolerant, montane to subalpine, circumpolar liverworts (transcontinental in North America). Both species occur in subalpine and montane boreal climates on moderately dry to fresh, nitrogen-poor soils. Plentiful on thick, snow-compacted forest floors (often on decaying wood) on water-shedding sites in coniferous forests. Usually associated with *Rhytidiopsis robusta* and *Vaccinium membranaceum*. An oxylophytic species characteristic of very acid Mor humus forms.

Bartramia pomiformis

Bartramiaceae

A shade-tolerant, submontane to alpine, cosmopolitan moss (transcontinental in North America). Grows on water-shedding sites within tundra, boreal, temperate, and mesothermal climates. Inhabits shaded crevices of base-rich boulders and rocks with very shallow accumulations of soils rich in organic matter. Characteristic of colluvial sites.

Bazzania tricrenata

Lepidoziaceae

A shade-tolerant, submontane to montane, circumpolar liverwort (distributed in Pacific, Cordilleran, and Atlantic North America). Occurs in hyper-maritime to maritime cool mesothermal climates on fresh to very moist, nitrogen-poor soils. Scattered to plentiful in coniferous forests on water-shedding and water-receiving sites; its occurrence decreases with increasing elevation and continentality. Often inhabits decaying wood and bark of coniferous trees, particularly in very wet climates. An oxylophytic species characteristic of Mor humus forms.

Blechnum spicant
Deer fern

Blechnaceae
(Deer fern family)

A shade-tolerant, submontane to alpine, circumpolar fern distributed more in Pacific than Cordilleran North America. Occurs in hypermaritime to maritime subalpine boreal and summer-wet cool mesothermal climates on fresh to very moist, nitrogen-poor soils. Its occurrence decreases with increasing continentality (very frost-sensitive). Scattered to abundant (occasionally dominant) in old-growth coniferous forests on water-receiving sites; sporadic and less vigorous on water-collecting sites. Grows best on well decomposed (greasy) organic materials; on nutrient-rich soils, confined to decaying coniferous wood. Usually associated with *Gaultheria shallon*, *Rhytidiadelphus loreus* and *Vaccinium alaskaense*. An oxylophytic species characteristic of Mor humus forms.

Boschniakia hookeri
Groundcone

Orobanchaceae
(Broomrape family)

A shade-tolerant/intolerant, submontane to montane, Pacific North American parasite. Occurs in maritime summer-dry cool mesothermal climates on nitrogen-poor soils. Its occurrence decreases with increasing latitude, elevation, and continentality. Sparse in open-canopy, Douglas-fir forests on water-shedding sites. Parasitic on *Gaultheria shallon*. An oxylophytic species characteristic of Mor humus forms.

Boykinia elata
Boykinia occidentalis
Coast boykinia

Saxifragaceae
(Saxifrage family)

A shade-tolerant, submontane to montane, Pacific North American forb. Occurs in hyper-maritime to maritime cool mesothermal climates on fresh to very moist, nitrogen-rich soils. Its occurrence increases with increasing precipitation, and decreases with increasing elevation and continentality. Scattered in coniferous forests, plentiful in broad-leaved forests on water-receiving sites with fast-moving groundwater. Most common on flooded alluvial terraces and stream-edge sites; often associated with *Ribes bracteosum*. A nitro-phytic species characteristic of Moder and Mull humus forms.

Brachythecium albicans

Brachytheciaceae

A shade-tolerant/intolerant, submontane to subalpine, European and transcontinental North American moss. Occurs on moderately dry to fresh, nutrient-medium soils within boreal, temperate, and mesothermal climates. Sporadic to scattered in grassy communities and open-canopy forests on water-shedding sites. Characteristic of young-seral forests.

Bromus carinatus
California brome

Poaceae
(Grass family)

A very shade-intolerant, montane to subalpine, Western North American grass distributed more in the Cordilleran than the Pacific region. Occurs predominantly in continental climates on very dry to moderately dry, nitrogen-medium soils. Its occurrence increases with increasing temperature and continentality, and decreases with increasing precipitation. Scattered on water-shedding sites within boreal, temperate, cool semiarid, and marginally summer-dry mesothermal climates. Occasional in grassy communities and open-canopy, young-seral forests. Characteristic of moisture-deficient sites.

Bromus vulgaris
Columbia brome

Poaceae
(Grass family)

A shade-tolerant/intolerant, submontane to montane, Western North American grass distributed equally in the Pacific and Cordilleran regions. Occurs on nitrogen-rich soils within montane, boreal, wet temperate, and cool mesothermal climates; its occurrence decreases with elevation. Scattered in coniferous forests on water-shedding sites, common in broad-leaved forests on water-receiving (alluvial, floodplain, and seepage) sites. A nitrophytic species characteristic of Moder and Mull humus forms.

Calamagrostis canadensis
Calamagrostis langsdorfii
Bluejoint

Poaceae
(Grass family)

A shade-tolerant/intolerant, submontane to subalpine, circumpolar grass (transcontinental in North America). Occurs on very moist to wet, nitrogen-medium soils within boreal, wet temperate, and cool mesothermal climates. Common in semi-terrestrial communities on water-receiving and collecting sites (floodplains, fens, and marshes). Characteristic of wetlands.

Calamagrostis nutkaensis
Pacific reedgrass

Poaceae
(Grass family)

A shade-tolerant/intolerant, submontane to subalpine (in hypermaritime climates), Pacific North American grass. Occurs in hypermaritime to maritime summer-wet cool mesothermal climates on very moist to wet, nitrogen-medium soils; its occurrence decreases with increasing elevation and continentality. Scattered to abundant on water-shedding and water-receiving sites in ocean-exposed forests of west Vancouver Island, central mainland, and Queen Charlotte Islands. Also inhabits non-forested, littoral sites (beaches, marshes, and rocky shores) that are affected by ocean spray, fog, and/or brackish water. Characteristic of hypermaritime forests.

Calamagrostis rubescens
Pinegrass, purple reedgrass

Poaceae
(Grass family)

A shade-tolerant/intolerant, montane, Asian and Western North American grass distributed in the Pacific (only in south), Cordilleran, and Central regions. Occurs in continental boreal and cool temperate climates on very dry to moderately dry, nitrogen-medium soils; its occurrence decreases with elevation and precipitation. Occasional in lodgepole pine and Douglas-fir forests on water-shedding sites in the southern coast-interior ecotone. Commonly associated with *Spiraea betulifolia*. Characteristic of continental forests.

Caltha biflora
Caltha leptosepala var. *biflora*
Two-flowered white marsh-marigold
Caltha leptosepala
Alpine white marsh-marigold

Ranunculaceae
(Buttercup family)

C. biflora - a shade-intolerant, montane to subalpine, Western North American forb distributed more in the Pacific than the Cordilleran region. Occurs in hypermaritime to maritime subalpine boreal and cool mesothermal climates. *C. leptosepala* - a very shade-intolerant, subalpine to alpine, Western North American forb distributed equally in the Pacific and Cordilleran regions. Occurs in alpine tundra and subalpine boreal climates.

Both species occur on water-receiving sites, on moist to wet, nutrient-rich soils. Scattered to abundant in herbaceous communities along small streams fed by melting snowbanks. Characteristic of subalpine communities.

Calypogeia trichomanis

Calypogeiaceae

A shade-tolerant, submontane to subalpine, circumpolar liverwort (distributed in Pacific, Cordilleran, and Atlantic North America). Occurs on fresh to very moist, nitrogen-poor soils within boreal, temperate, and mesothermal climates. Plentiful in coniferous forests on water-shedding and water-receiving sites; often inhabits decaying coniferous wood and silicaceous sand, gravel, or rocks. Usually associated with *Lepidozia reptans*, *Rhizomnium glabrescens*, *Rhytidiadelphus loreus*, and *Scapania bolanderi*. An oxylophytic species characteristic of Mor humus forms.

Calypso bulbosa
Fairy-slipper

Orchidaceae
(Orchid family)

A shade-tolerant/intolerant, submontane to montane, circumpolar forb (transcontinental in North America). Occurs on moderately dry to fresh, nitrogen-medium soils within montane boreal, cool temperate, and cool mesothermal climates. Rare to occasional (locally abundant on eastern Vancouver Island and Queen Charlotte Islands) in mossy forest understories on water-shedding sites; its occurrence increases with increasing continentality. Characteristic of coniferous forests.

Camassia quamash
Common camas
Camassia leichtlinii
Great camas

Liliaceae
(Lily family)

Shade-intolerant, submontane to montane, Western North American forbs distributed more in the Pacific than the Cordilleran region. Both species occur in maritime to submaritime summer-dry cool mesothermal climates on moderately dry to fresh, nitrogen-rich soils (Moder and Mull humus forms). Scattered to plentiful in open-canopy Garry oak stands on water-shedding sites; their occurrence decreases with increasing elevation and precipitation. Also inhabit meadow-like communities where early spring moisture is followed by mid-summer drought; occasionally found around vernal pools, springs, and intermittent streams.

Campanula scouleri
Scouler's harebell, Scouler's bluebell

Campanulaceae
(Hairbell family)

A shade-intolerant, submontane to montane, Western North American forb distributed more in the Pacific than the Cordilleran region. Occurs in cool mesothermal climates on very dry to moderately dry, nitrogen-poor soils (Mor humus forms). Sporadic in open-canopy forests on strongly drained, rocky and stony, water-shedding sites; also in non-forested communities on disturbed sites. An oxylophytic species characteristic of moisture-deficient sites.

Cardamine breweri
Brewer's bitter-cress

Brassicaceae
(Mustard family)

A shade-tolerant/intolerant, submontane to subalpine, Western North American forb distributed more in the Pacific than the Cordilleran region. Occurs in cool temperate and cool mesothermal climates on wet to very wet, nitrogen-rich soils (Moder and Mull humus forms); its occurrence decreases with increasing elevation. Inhabits depressions on water-collecting sites; with groundwater table at or above the ground surface. Scattered in open-canopy red alder, Sitka spruce, or western redcedar stands on gleysolic or organic soils affected by slow-moving seepage. Usually associated with *Athyrium filix-femina* and *Lysichitum americanum*. A nitrohphytic species characteristic of nutrient-rich wetlands.

Cardamine nuttallii
Cardamine pulcherrima
Dentaria tenella
Nuttall's bitter-cress

Brassicaceae
(Mustard family)

A shade-tolerant, submontane to montane, Pacific North American forb; ecologically comparable to *C. breweri*. Occurs in maritime to hypermaritime cool mesothermal climates on very moist to wet, nitrogen-rich soils; its occurrence decreases with increasing elevation and continentality. Sporadic in closed-canopy forests on water-receiving sites; occasional in depressions on water-collecting sites with groundwater table at the ground surface. A nitrophytic species characteristic of Moder or Mull humus forms.

Carex anthoxanthea
Sweet sedge

Cyperaceae
(Sedge family)

A shade-intolerant, submontane to subalpine, Asian and Western North American sedge distributed more in the Pacific than the Cordilleran region. Occurs in boreal and hypermaritime to maritime cool mesothermal climates on wet to very wet, nutrient-medium soils. Its occurrence decreases with increasing elevation and continentality. Common in non-forested, semi-terrestrial communities on water-collecting sites, occasional in open-canopy forests on seepage sites. Characteristic of wet forests and wetlands.

Carex deweyana
Carex bolanderi , Carex leptopoda
Dewey's sedge

Cyperaceae
(Sedge family)

A shade-tolerant/intolerant, submontane to montane, Asian and North American sedge distributed in the Pacific and Cordilleran regions. Occurs on fresh to very moist, nitrogen-rich soils, often with a fluctuating groundwater table. Widespread in boreal, cool temperate, and cool mesothermal climates; its occurrence decreases with elevation. Scattered on water-receiving and water-collecting sites; most common on floodplain and seepage sites in fern-dominated forest understories; on the southern coast often grows with *Carex hendersonii*. A nitrophytic species characteristic of Moder and Mull humus forms.

Carex hendersonii
Henderson's sedge

Cyperaceae
(Sedge family)

A shade-tolerant/intolerant, submontane to montane, Western North American sedge distributed more in the Pacific than the Cordilleran region. Occurs in maritime to submaritime cool mesothermal climates on very moist to wet, nitrogen-rich soils, often with a fluctuating groundwater table. Sporadic to abundant (in broad-leaved stands) on water-receiving and water-collecting sites (edaphically comparable to *Carex deweyana*); its occurrence decreases with elevation, latitude, and continentality. Usually associated with *Polystichum munitum* and *Tiarella trifoliata*. A nitrophytic species characteristic of Moder and Mull humus forms.

Carex inops
Carex pensylvanica var *vespertina*
Long-stoloned sedge

Cyperaceae
(Sedge family)

A shade-intolerant, submontane to montane, Pacific North American sedge. Occurs in maritime to submaritime summer-dry cool mesothermal climates on moderately dry to fresh, nitrogen-medium soils. Sporadic on water-shedding, often disturbed, sites. Characteristic of grassy communities and open-canopy, young-seral, Douglas-fir forests.

97

Carex laeviculmis
Smooth-stemmed sedge

Cyperaceae
(Sedge family)

A shade-tolerant/intolerant, submontane to subalpine, Asian and Western North American sedge distributed equally in the Pacific and Cordilleran regions. Occurs on wet to very wet, nitrogen-medium soils within boreal, temperate, and cool mesothermal climates. Scattered in non-forested, semi-terrestrial communities (less frequent in forest understories) on water-collecting sites; often with *Lysichitum americanum*. Characteristic of wetlands.

Carex livida
Pale sedge

Cyperaceae
(Sedge family)

A very shade-intolerant, submontane to montane, circumpolar sedge (transcontinental in North America). Occurs on wet to very wet, nitrogen-medium soils within boreal, cool temperate, and cool mesothermal climates. Scattered in semi-terrestrial communities on gleysolic or organic soils with slow-moving groundwater near or at the ground surface (predominantly in fens, bogs, and marshes). Characteristic of wetlands.

Carex mertensii
Merten's sedge

Cyperaceae
(Sedge family)

A shade-intolerant, montane to subalpine, Asian and Western North American sedge distributed more in the Pacific than the Cordilleran region. Occurs on fresh to very moist, nitrogen-rich soils within boreal, humid cool temperate, and cool mesothermal climates. Inhabits exposed mineral soil on high-elevation clearcuts, clearings, and waysides; often near streams or on seepage sites. A nitrophytic species characteristic of disturbed sites.

Carex obnupta
Slough sedge

Cyperaceae
(Sedge family)

A shade-tolerant/intolerant, submontane to montane, Pacific North American sedge. Occurs in hypermaritime to submaritime cool mesothermal climates on wet to very wet, nitrogen-rich soils (Moder and Mull humus forms); its occurrence decreases with increasing elevation and continentality. Scattered to plentiful (often dominant) in graminoid-dominated, semi-terrestrial communities; on water-collecting sites (fens, bogs, and marshes) with gleysolic or organic soils. A nitrophytic species characteristic of nutrient-rich wetlands.

Carex rossii
Carex deflexa
Ross' sedge

Cyperaceae
(Sedge family)

A shade-intolerant, submontane to subalpine, North American sedge distributed in the Pacific, Cordilleran, and Central regions. Occurs on very dry to moderately dry, nitrogen-medium soils, within boreal, temperate, cool semi-arid, and cool mesothermal climates. Its occurrence increases with increasing continentality and decreases with increasing precipitation. Sporadic on shallow stony or rocky soils in open-canopy forests; scattered on cutover areas on water-shedding sites. Characteristic of moisture-deficient sites.

Carex sitchensis
Sitka sedge

Cyperaceae
(Sedge family)

A very shade-intolerant, submontane to montane, Western North American sedge distributed more in the Pacific than the Cordilleran region. Occurs in cool mesothermal climates on wet to very wet, nitrogen-rich soils (Moder and Mull humus forms); its occurrence decreases with increasing elevation and continentality. Scattered to plentiful in semi-terrestrial communities on water-collecting sites (fresh and brackish water marshes) with gleysolic or organic soils and with more or less stagnant water tables at or above the ground surface. Often associated with *Carex obnupta*. Characteristic of nutrient-rich wetlands.

Cassiope mertensiana
White mountain-heather

Ericaceae
(Heath family)

A very shade-intolerant, subalpine to alpine, Western North American evergreen shrub; distributed more in the Pacific than the Cordilleran region. Occurs in maritime to subcontinental alpine tundra and subalpine boreal climates on moderately dry to fresh, nitrogen-poor soils (Mor humus forms). Plentiful to abundant in heath communities on water-shedding sites; its occurrence decreases with increasing continentality. Associated with *Barbilophozia floerkei, Cassiope stelleriana, Phyllodoce empetriformis,* and *Vaccinium deliciosum.* An oxylophytic species characteristic of alpine communities.

Cassiope stelleriana
Alaskan mountain-heather

Ericaceae
(Heath family)

A very shade-intolerant, sub-alpine to alpine, Asian and Western North American ever-green shrub distributed more in the Pacific than the Cordilleran region. Occurs in maritime to submaritime alpine tundra and subalpine boreal climates on fresh to very moist, nitrogen-poor soils (Mor humus forms). Inhabits water-shedding and water-receiving sites that are often affected by seepage from snowbanks. Plentiful to abundant in heath communities; its occurrence decreases with increasing continentality. Associated with *Barbilophozia floerkei, Cassiope mertensiana, Phyllodoce empetriformis,* and *Vaccinium deliciosum.* An oxylophytic species characteristic of alpine communities.

Cassiope tetragona
Four-angled mountain-heather

Ericaceae
(Heath family)

A very shade-intolerant, alpine, circumpolar, evergreen shrub (transcontinental in North America). Occurs in continental alpine tundra and subalpine boreal climates on fresh to very moist, nitrogen-medium soils. Scattered to plentiful in heath communities affected by snow-drifts; its occurrence increases with increasing continentality. Often associated with *Cassiope mertensiana* and *Phyllodoce glanduliflora* in the coast-interior ecotone. Characteristic of alpine communities.

Ceanothus sanguineus
Redstem ceanothus

Rhamnaceae
(Buckthorn family)

A very shade-intolerant, sub-montane to montane, Western North American deciduous shrub distributed more in the Cordilleran than the Pacific region. Occurs in continental climates on very dry to moderately dry, nitrogen-medium soils. Scattered and often plentiful on disturbed water-shedding sites within cool temperate, cool semiarid, and dry cool mesothermal climates; its occurrence increases with continentality. Common in early seral communities in the coastal-interior ecotone. When forming a shrub layer it hinders regeneration and growth of shade-intolerant conifers. Symbiosis with nitrogen-fixing bacteria enhances the supply of available soil nitrogen. Characteristic of moisture-deficient sites.

Ceanothus velutinus
Snowbrush ceanothus

Rhamnaceae
(Buckthorn family)

A shade-intolerant, submontane to subalpine, Cordilleran North American evergreen shrub. Occurs in continental cool temperate and cool semiarid climates on moderately dry to fresh, nitrogen-medium soils, sites. Scattered to abundant in early-seral communities on disturbed, water-shedding sites; its occurrence increases with increasing continentality (occasional in the coastal-interior ecotone). When forming a shrub layer, it hinders natural regeneration and growth of shade-intolerant conifers. Symbiosis with nitrogen-fixing bacteria enhances the supply of available soil nitrogen. Characteristic of continental forests.

Ceratodon purpureus

Ditrichaceae

A shade-intolerant, submontane to alpine, cosmopolitan moss (transcontinental in North America). Grows on a wide range of sites within tundra, boreal, temperate, and mesothermal climates. Inhabits a variety of substrates including coarse fragments, decaying wood, and mineral soil (most often occurs on exposed mineral soil). Characteristic of fire-disturbed sites.

Chamaecyparis nootkatensis
Yellow-cedar

Cupressaceae
(Cypress family)

A shade-tolerant/intolerant, submontane to subalpine, Western North American evergreen conifer distributed more in the Pacific than the Cordilleran region. Occurs in hypermaritime to submaritime subalpine boreal and summer-wet cool mesothermal climates; its occurrence increases with increasing elevation and decreases with increasing continentality. Occasionally forms pure stands on wet sites, sporadic to plentiful in mixed-species stands (usually with Pacific silver fir, western redcedar, western hemlock, or mountain hemlock) on a variety of sites. As does western redcedar, yellow-cedar tolerates a nearly complete soil range, and develops a very dense root system; the latter may explain its abundance on very steep, seepage-affected, often unstable, colluvial slopes. Most productive on montane, moist, nutrient-rich (seepage) sites within wet cool mesothermal climates. Characteristic of maritime subalpine forests.

Chimaphila menziesii
Menzies' pipsissewa

Pyrolaceae
(Wintergreen family)

A shade-tolerant, submontane to montane, Western North American evergreen shrub distributed more in the Pacific than the Cordilleran region. Occurs in maritime to submaritime cool mesothermal climates on moderately dry to fresh, nitrogen-medium soils; its occurrence decreases with increasing elevation and continentality. Sporadic in coniferous stands on water-shedding sites (less often on water-receiving sites). Usually associated with *Hylocomium splendens*, *Kindbergia oregana*, *Rhytidiadelphus loreus*, and *Rhytidiopsis robusta*. Characteristic of mesothermal coniferous forests.

Chimaphila umbellata
Prince's pine

Pyrolaceae
(Wintergreen family)

A shade-tolerant/intolerant, submontane to montane, circumpolar evergreen shrub (transcontinental in North America). Occurs on very dry to moderately dry, nitrogen-poor soils (Mor humus forms) within montane boreal, temperate, and cool mesothermal climates. Its occurrence decreases with increasing elevation and precipitation, and increases with increasing continentality. Sporadic to scattered in coniferous forests on water-shedding sites with coarse-skeletal soils. Typically associated with *Hylocomium splendens*. An oxylophytic species characteristic of moisture-deficient sites.

Cinna latifolia
Nodding wood-reed

Poaceae
(Grass family)

A shade-tolerant, submontane to subalpine, circumpolar grass (transcontinental in North America). Occurs on fresh to very moist, nitrogen-rich soils within boreal, wet temperate, and cool mesothermal climates. Tolerates fluctuating groundwater tables. Sporadic to scattered in the herbaceous understory of broad-leaved forests on water-receiving (alluvial, floodplain, seepage, and stream-edge) sites. Often inhabits disturbed microsites on and in recent cut-over areas. A nitrophytic species characteristic of Moder and Mull humus forms.

Circaea alpina
Alpine enchanter's nightshade

Onagraceae
(Evening-Primrose family)

A shade-tolerant, submontane to subalpine, circumpolar forb (transcontinental in North America). Occurs in boreal, cool temperate, and cool mesothermal climates on fresh to very moist, nitrogen-medium soils. Sporadic in coniferous forests on water-receiving sites; its occurrence increases with increasing continentality. Often on decaying wood; occasional in nutrient-poor wetlands. Frequently associated with *Athyrium filix-femina* and *Oplopanax horridus*. Characteristic of moist forests.

Circaea pacifica
Circaea alpina ssp. *pacifica*
Pacific enchanter's nightshade

Onagraceae
(Evening-Primrose family)

A shade-tolerant/intolerant, submontane to montane, Western North American forb distributed more in the Pacific than the Cordilleran region. Occurs in maritime to submaritime cool mesothermal climates on fresh to very moist, nitrogen-rich soils, often with a fluctuating groundwater table. Its occurrence decreases with increasing elevation, precipitation, and continentality. Common on water-receiving (flooded) sites in the herbaceous understory of broad-leaved forests (black cottonwood, broad-leaved maple, or red alder); often inhabits exposed mineral soil. A nitrophytic species characteristic of Moder and Mull humus forms.

Cladina mitis
Cladina, Cladonia spp.

Cladoniaceae

Shade-intolerant, submontane to alpine, circumpolar lichens (transcontinental in North America). These lichens occur on excessively dry to very dry, nitrogen-poor soils within tundra, boreal, cool temperate, and cool mesothermal climates. Scattered to plentiful in the open or in open-canopy forests on strongly drained, water-shedding sites with shallow and/or coarse-skeletal soils.

Inhabit very thin, acid organic substrates (Mor humus forms), less frequently coarse fragments and decaying wood. Occasional in nutrient-poor wetlands on topographic prominences subjected to regular desiccation.

Cladina arbuscula, C. impexa, C. rangiferina, Cladonia bellidiflora, C. gracilis and other species of *Cladina* and *Cladonia* are all oxylophytic species characteristic of moisture-deficient sites.

Cladothamnus pyroliflorus
Copperbush

Ericaceae
(Heath family)

A shade-tolerant/intolerant, montane to subalpine, Pacific North American deciduous shrub. Occurs in maritime to submaritime subalpine boreal climates on nitrogen-poor soils; its occurrence decreases with increasing continentality. Grows in coniferous forests (mountain hemlock, Pacific silver fir, and yellow-cedar) on water-shedding or water-receiving sites. Typically associated with *Rhytidiopsis robusta, Vaccinium alaskaense, V. membranaceum,* and *V. ovalifolium.* An oxylophytic species characteristic of Mor humus forms.

Claopodium crispifolium   *Thuidiaceae*

A shade-tolerant, submontane to montane, Asian and Pacific North American moss; its occurrence decreases with increasing continentality. Occurs in cool temperate and cool mesothermal climates on very shallow, calcium-rich soils. Inhabits coarse fragments, cliffs, or bark of trees in shaded forest understories. Characteristic of calcium-rich substrates.

Claytonia sibirica   *Portulacaceae*
Montia sibirica   (Purslane family)
Siberian miner's-lettuce

A shade-tolerant, submontane to montane, Asian and Western North American forb distributed more in the Pacific than the Cordilleran region. Occurs in cool temperate and cool mesothermal climates on fresh to very moist, nitrogen-rich soils, often with a fluctuating groundwater table. Its occurrence decreases with increasing elevation and precipitation. Common on water-receiving sites; most frequent in broad-leaved forests on alluvial, floodplain, seepage, and stream-edge sites. A nitrophytic species characteristic of Moder and Mull humus forms.

Clintonia uniflora
Queen's cup

Liliaceae
(Lily family)

A very shade-tolerant, montane to subalpine, Western North American forb distributed more in the Pacific than the Cordilleran region. Occurs in boreal and cool temperate climates on moderately dry to fresh, nitrogen-poor soils. Common in the shaded understory of coniferous forests on water-shedding and water-receiving sites. Typically associated with *Hylocomium splendens*, *Pleurozium schreberi*, *Ptilium crista-castrensis*, *Rhytidiopsis robusta*, *Rubus pedatus*, and *Vaccinium membranaceum.* An oxylophytic species characteristic of Mor humus forms.

Collinsia parviflora
Small-flowered blue-eyed Mary

Scrophulariaceae
(Figwort family)

A shade-intolerant, submontane to montane, North American forb distributed in Pacific, Cordilleran, and Central regions. Occurs on very dry to moderately dry, nitrogen-medium soils within boreal, temperate, cool semiarid, and mesothermal climates. Its occurrence increases with increasing temperature and decreases with increasing precipitation. Occasional in the open and in open-canopy forests on very shallow soils on rock outcrops and cliffs. Often inhabits meadow-like communities on water shedding-sites where early spring moisture is followed by mid-summer drought. Characteristic of moisture-deficient sites.

Conocephalum conicum

Conocephalaceae

A shade-tolerant, submontane to montane, circumpolar liverwort; (transcontinental in North America). Occurs on very moist to wet, calcium-rich and nitrogen-rich soils within boreal, temperate, and mesothermal climates; its occurrence decreases with increasing elevation. Inhabits a variety of calcium-rich substrates in shaded forest understories on water-receiving and water-collecting sites: friable forest floors, exposed and often melanized mineral soils, base-rich coarse fragments, and rocks affected by surface seepage; most common near streams and springs. Characteristic of calcium-rich substrates.

Coptis aspleniifolia
Fern-leaved goldthread
Coptis trifolia
Three-leaved goldthread

Ranunculaceae
(Buttercup family)

C. aspleniifolia - a shade-tolerant/intolerant, submontane to subalpine, Pacific North American forb. Occurs in hypermaritime to maritime subalpine boreal and cool mesothermal climates on fresh to very moist, nitrogen-poor soils; its occurrence decreases with increasing continentality. *C. trifolia* - a shade-intolerant, submontane to montane, Asian and transcontinental North American forb. Occurs on very moist to wet, nitrogen-poor soils within boreal cool temperate and cool mesothermal climates.

Scattered in open-canopy coniferous forests on gleysolic or organic soils on water-receiving sites; *C. trifolia* common in nutrient-poor wetlands. Oxylophytic species characteristic of Mor humus forms.

Corallorhiza maculata
Spotted coralroot
Corallorhiza mertensiana
Western coralroot

Orchidaceae
(Orchid family)

Shade-tolerant, submontane to subalpine, saprophytes (*C. maculata* - transcontinental North American; *C. mertensiana* - Western North American). Both species occur on moderately dry to fresh, nitrogen-poor soils within boreal, temperate, and cool mesothermal climates. Sporadic in the mossy understory of coniferous forests on water-shedding and water-receiving sites; commonly associated with *Gaultheria shallon*, *Hylocomium splendens*, and *Rhytidiadelphus loreus*. Oxylophytic species characteristic of Mor humus forms.

Cornus unalaschkensis
Cordilleran bunchberry
Cornus canadensis
Bunchberry

Cornaceae
(Dogwood family)

C. unalaschkensis - a shade-tolerant/intolerant, submontane to montane, Pacific North American forb. Occurs in hypermaritime to maritime subalpine boreal and summer-wet cool mesothermal climates. *C. canadensis* - a shade-tolerant, submontane to subalpine, Asian and transcontinental North American forb. Occurs in boreal and cool temperate climates.

Scattered to plentiful in the mossy understory of coniferous forests on water-shedding and water-receiving sites, on nitrogen-poor soils. Often inhabit decaying wood, sides of large stumps and tree trunks, and topographic prominences in nutrient-poor wetlands. Oxylophytic species characteristic of Mor humus forms.

Cornus nuttallii
Western flowering dogwood

Cornaceae
(Dogwood family)

A shade-tolerant/intolerant, submontane to montane, Western North American deciduous broad-leaved tree distributed more in the Pacific than the Cordilleran region. Occurs in maritime to submaritime cool mesothermal climates on moderately dry to fresh, nitrogen-rich soils (Moder or Mull humus forms). A shrub or small tree in disturbed communities and coniferous forests on water-shedding sites, most often on colluvial slopes. Its occurrence decreases with increasing latitude, precipitation, and continentality. Characteristic of young-seral mesothermal forests.

Cornus sericea
Red-osier dogwood

Cornaceae
(Dogwood family)

A shade-tolerant/intolerant, submontane to subalpine, transcontinental North American deciduous shrub. Occurs on very moist to wet, nitrogen-rich soils (Moder and Mull humus forms. Tolerates fluctuating groundwater tables. Widespread in boreal, temperate, and cool mesothermal climates; its occurrence decreases with increasing elevation. Plentiful in black cottonwood or red alder stands on alluvial, flooded, and stream-edge sites; often dominant in nutrient-rich wetlands. Regenerates abundantly and grows vigorously in clearings where it may hinder natural regeneration and growth of shade-intolerant conifers. Characteristic of alluvial floodplain forests.

Corylus cornuta
Beaked hazelnut

Betulaceae
(Birch family)

A shade-tolerant/intolerant, submontane to montane, transcontinental North American deciduous shrub. Occurs in cool temperate and cool mesothermal climates on moderately dry to fresh, calcium-rich and nitrogen-rich soils (Moder and Mull humus forms). Sporadic in disturbed forests on water-shedding sites; its occurrence decreases with increasing latitude and precipitation, and increases with increasing continentality. Characteristic of young-seral broad-leaved forests.

Crataegus douglasii
Black hawthorn

Rosaceae
(Rose family)

A shade-tolerant/intolerant, submontane to montane, North American deciduous shrub distributed in the Pacific, Cordilleran, and Central regions. Occurs on very moist to wet, nitrogen-rich soils (Moder and Mull humus forms) within boreal, temperate, cool semiarid, and cool mesothermal climates. Sporadic in open-canopy forests on water-receiving sites, scattered in poorly forested, semi-terrestrial communities on water-collecting sites (swamps). Characteristic of nutrient-rich wetlands.

Cryptogramma crispa
Cryptogramma acrostichoides
Parsley fern

Adiantaceae
(Maidenhair fern family)

A shade-intolerant, submontane to subalpine, circumpolar fern distributed equally in Pacific, Cordilleran, and Central North America. Occurs on very dry to moderately dry, nitrogen-poor soils (Mor humus forms) within boreal, cool temperate, and cool mesothermal climates. Occurs sporadically on exposed, strongly drained treeless rock outcrops or colluvial soils where it inhabits microsites with very shallow, organic matter-rich soils. An oxylophytic species characteristic of moisture-deficient sites.

Cystopteris fragilis
Fragile fern

Aspleniaceae
(Spleenwort family)

A shade-tolerant/intolerant, submontane to alpine, cosmopolitan fern (transcontinental in North America). Occurs on fresh to very moist, nitrogen-medium soils within tundra, boreal, temperate, and mesothermal climates. Rare to sporadic in non-forested communities and semi-open forests on water-shedding or water-receiving sites. Characteristic of colluvial sites.

Cytisus scoparius
Sarothamnus scoparius
Scotch broom

Fabaceae
(Pea family)

A shade-intolerant, submontane to montane, European deciduous shrub introduced to Pacific and Atlantic North America. Occurs in maritime to submaritime summer-dry cool mesothermal climates on very dry to moderately dry, nitrogen-medium soils. Its occurrence decreases with increasing latitude, elevation, and continentality. Inhabits exposed mineral soil in early-seral, non-forested communities on strongly drained, water-shedding sites. Symbiotic with nitrogen-fixing organisms. Characteristic of disturbed sites.

Danthonia intermedia
Timber oatgrass
Danthonia spicata
Poverty oatgrass

Poaceae
(Grass family)

D. intermedia - a shade-intolerant, submontane to subalpine, Asian and trans-continental North American grass. *D. spicata* - a shade-intolerant, submontane to montane, transcontinental North American grass.

Both species occur on very dry to moderately dry, nitrogen-poor soils (Mor humus forms) within boreal, temperate, and cool mesothermal climates; their occurrence decreases with increasing elevation and precipitation. Sporadic in open-canopy, young-seral forests on strongly drained and shallow soils on water-shedding sites; often present in non-forested communities on rock outcrops. Characteristic of moisture-deficient sites.

Deschampsia cespitosa
Tufted hairgrass

Poaceae
(Grass family)

A shade-intolerant, submontane to subalpine, circumpolar grass (transcontinental in North America). Occurs on very moist to wet, calcium-rich and nitrogen-rich soils (Moder and Mull humus forms), often with a fluctuating groundwater table. Grows in boreal, temperate, and cool mesothermal climates on water-receiving and water-collecting sites with slow-moving, often stagnant groundwater. Frequent on disturbed gleysolic soils or tidal marshes (tolerates brackish water and ocean spray). Characteristic of early-seral or semi-terrestrial, non-forested communities.

Dicentra formosa
Bleeding heart

Fumariaceae
(Fumitory family)

A shade-tolerant/intolerant, submontane to montane, Western North American forb distributed more in the Pacific than the Cordilleran region. Occurs in maritime to submaritime cool mesothermal climates on fresh to very moist, nitrogen-rich soils; its occurrence decreases with increasing elevation and continentality. Sporadic to plentiful in the herbaceous understory of young-seral forests on water-receiving sites; most common in broad-leaved forests. Occasional in early-seral communities on disturbed sites (burns and clearings). A nitrophytic species characteristic of Moder and Mull humus forms.

Dicranum fuscescens   Dicranaceae
Dicranum tauricum

Shade-tolerant, submontane to subalpine mosses (*D. fuscescens* - circumpolar, transcontinental in North America; *D. tauricum* - European and Western North American).

Both species occur on very dry to moderately dry, nitrogen-poor soils within boreal, temperate, and mesothermal climates. Occasional to plentiful in coniferous forests on water-shedding sites; inhabit forest floor on decaying wood, occasional on the bark of trees. Oxylophytic species characteristic of Mor humus forms.

Dicranum howellii   *Dicranaceae*
Dicranum pallidisetum

Shade-tolerant Western North American mosses distributed more in the Pacific than the Cordilleran region. *D. howellii* - a submontane to subalpine species which grows in boreal, temperate, and mesothermal climates on water-shedding sites; its occurrence decreases with increasing elevation and continentality. *D. pallidisetum* - a subalpine to alpine species occurring more in continental alpine tundra and subalpine boreal climates.

Plentiful to abundant on moderately dry to fresh, nitrogen-poor soils in coniferous forests and open areas; very often inhabit decaying wood. Oxylophytic species characteristic of Mor humus forms.

Disporum hookeri
Hooker's fairybells

Liliaceae
(Lily family)

A shade-tolerant, submontane to montane, Western North American forb distributed equally in the Pacific and Cordilleran regions. Occurs on fresh to very moist, nitrogen-rich soils within montane boreal, cool temperate, and cool mesothermal climates. Its occurrence decreases with increasing latitude and elevation. Scattered on water-shedding sites; common in broad-leaved forests on water-receiving (alluvial, floodplain, colluvial, and seepage) sites. A nitrophytic species characteristic of Moder and Mull humus forms.

Disporum smithii
Smith's fairybells

Liliaceae
(Lily family)

A shade-tolerant, submontane to montane, Pacific North American forb; ecologically comparable to *D. hookeri*. Occurs in maritime to submaritime, cool mesothermal climates on fresh to very moist, nitrogen-rich soils. Very sparse in herbaceous forest understories on water-receiving sites; its occurrence decreases with increasing latitude, elevation, and continentality. Characteristic of Moder and Mull humus forms.

Disporum trachycarpum
Rough-fruited fairybells

Liliaceae
(Lily family)

A shade-tolerant/intolerant, montane, North American forb distributed in the Cordilleran and Central regions. Occurs in continental cool temperate climates on nitrogen-rich soils (Moder or Mull humus forms). Scattered in the herbaceous understory of open-canopy forests on water-shedding and water-receiving sites in the coast-interior ecotone. Its occurrence increases with increasing continentality and decreases with increasing latitude. Characteristic of continental temperate forests.

Dodecatheon hendersonii
Broad-leaved shootingstar

Primulaceae
(Primrose family)

A shade-tolerant/intolerant, submontane to montane, Western North American forb distributed more in the Pacific than the Cordilleran region. Occurs in maritime to submaritime cool mesothermal climates on moderately dry to fresh, nitrogen-rich soils. Sporadic in open-canopy forests on water-shedding sites, or in meadow-like communities where early spring moisture is followed by mid-summer drought. Occasionally inhabits water-receiving sites (vernal springs). Its occurrence decreases with increasing latitude, elevation, precipitation, and continentality. Characteristic of Moder and Mull humus forms.

Dodecatheon pulchellum
Dodecatheon pauciflorum
Few-flowered shootingstar

Primulaceae
(Primrose family)

A very shade-intolerant, submontane to sub-alpine, North American forb distributed in the Pacific, Cordilleran, and Central regions. Occurs on moderately dry to fresh circumneutral to weakly alkaline, nitrogen-rich soils within alpine tundra, boreal, temperate, cool semiarid, and cool mesothermal climates. Scattered on water-shedding sites, occasional on water-receiving sites (vernal springs). Often found in meadow-like communities where early spring moisture is followed by midsummer drought. Characteristic of Moder and Mull humus forms.

Drosera rotundifolia
Round-leaved sundew

Droseraceae
(Sundew family)

A very shade-intolerant, submontane to sub-alpine, circumpolar forb (transcontinental in North America). Occurs on wet to very wet, nitrogen-poor soils (Mor humus forms) within boreal, cool temperate, and cool mesothermal climates. Sporadic in non-forested, semi-terrestrial communities on water-collecting sites amidst *Sphagnum* species in peat bogs. An oxylophytic species characteristic of nutrient-poor wetlands.

Dryopteris expansa
Dryopteris assimilis
Dryopteris austriaca
Dryopteris spinulosa
Spiny wood fern

Aspleniaceae
(Spleenwort family)

A shade-tolerant, submontane to subalpine, circumpolar fern (transcontinental in North America). Occurs on fresh to very moist, nitrogen-medium soils within boreal, cool temperate, and cool mesothermal climates; its occurrence increases with increasing precipitation. Common and occasionally dominant in coniferous forests on water-shedding and water-receiving sites; often inhabits decaying wood. Associated with *Blechnum spicant, Polystichum munitum, Rhytidiadelphus loreus*, and *Vaccinium alaskaense*. Characteristic of friable Mor and acidic Moder humus forms.

Dryopteris filix-mas
Male fern

Aspleniaceae
(Spleenwort family)

A shade-tolerant/intolerant, montane, circumpolar fern (transcontinental in North America). Occurs on fresh to very moist, nitrogen-rich soils within boreal, temperate, and cool mesothermal climates. Sporadic on water-shedding and water-receiving sites, alluvial forests and avalanche tracks. Typically associated with *Athyrium filix-femina, Gymnocarpium dryopteris, Oplopanax horridus*, and *Tiarella unifoliata*. Characteristic of Moder and Mull humus forms.

Elymus glaucus
Blue wildrye

Poaceae
(Grass family)

A shade-tolerant/intolerant, submontane to montane, Western North American grass distributed in the Pacific and Cordilleran regions (introduced to Eastern and Central North America). Occurs in cool temperate and cool mesothermal climates on moderately dry to fresh, nitrogen-rich soils. Sporadic on water-shedding sites, more frequent in broad-leaved forests on water-receiving (floodplain and stream-edge) sites. Its occurrence decreases with increasing precipitation and elevation. Characteristic of Moder and Mull humus forms.

Elymus hirsutus
Hairy wildrye

Poaceae
(Grass family)

A shade-tolerant/intolerant, submontane to subalpine, Western North American grass distributed more in the Pacific than the Cordilleran region. Occurs on very moist to wet, nitrogen-rich soils within boreal, cool temperate, and cool mesothermal climates; its occurrence decreases with increasing continentality. Sporadic in coniferous forests on water-receiving sites, more frequent in broad-leaved forests on floodplains. Characteristic of Moder and Mull humus forms.

Empetrum nigrum
Crowberry

Empetraceae
(Crowberry family)

A shade-intolerant, submontane to alpine, circumpolar,
evergreen shrub (transcontinental in North America).
Grows on a wide range of sites
in tundra, boreal, and cool
mesothermal climates. Most
often in nitrogen-poor soils in
semi-terrestrial communities
(peat bogs) where it inhabits
topographic prominences. An
oxylophytic species characteristic of Mor humus forms.

Epilobium angustifolium
Fireweed

Onagraceae
(Evening-primrose family)

A very shade-intolerant, submontane to subalpine, circumpolar forb (transcontinental in
North America). Grows on recently cutover
and/or burnt sites within boreal, temperate,
and cool mesothermal climates; its occurrence decreases with increasing elevation.
Scattered to abundant (often dominant) in
herbaceous communities on a wide range of
sites where it indicates increased decomposition of the remaining forest floor materials
(originally Mor humus forms). In some situations, high-density stands of this species
may hinder the survival and growth of forest
plantations. Commonly associated with *Polytrichum juniperinum*. A nitrophytic species
characteristic of fire-disturbed sites.

Epilobium latifolium
Broad-leaved willowherb

Onagraceae
(Evening-primrose family)

A very shade-intolerant, montane to alpine, circumpolar forb (transcontinental in North America). Occurs in tundra and boreal climates on nitrogen-rich soils. Scattered to plentiful in non-forested communities on water-receiving sites; its occurrence increases with increasing latitude. Commonly along intermittent streams; often on exposed mineral soil in early-seral communities. A nitrophytic species characteristic of disturbed sites.

Equisetum arvense
Common horsetail

Equisetaceae
(Horsetail family)

A shade-tolerant/intolerant, submontane to subalpine, circumpolar horsetail (transcontinental in North America). Grows on nitrogen-medium soils within tundra, boreal, temperate, and mesothermal climates. Common on water-receiving (floodplain, seepage, springs and ephemeral streams) sites, frequently dominant in early-seral communities and forest openings. Characteristic of disturbed sites.

Equisetum hyemale
Fall scouring-rush

Equisetaceae
(Horsetail family)

A shade-tolerant/intolerant, submontane to montane, circumpolar horsetail (transcontinental in North America). Occurs on fresh to moist, calcium-rich and nitrogen-rich soils (Moder or Mull humus forms) within boreal, temperate, cool semiarid, mesothermal, and tropical climates; its occurrence decreases with increasing elevation. Inhabits exposed mineral soil; scattered in broad-leaved forests on water-receiving (floodplain and stream-edge) sites, frequent in non-forested, early-seral communities. Characteristic of alluvial floodplain forests.

Equisetum sylvaticum
Wood horsetail

Equisetaceae
(Horsetail family)

A shade-tolerant, submontane to subalpine, circumpolar horsetail (transcontinental in North America, but present only in the northern part of the Pacific region). Occurs in continental boreal and cool temperate climates on moist to wet, nitrogen-poor soils. Common, often dominant, in coniferous forests on water-receiving sites with gleysolic and organic soils in the coast-interior ecotone. Occasional on flooded sites, in peat bogs, and on recent burns and clearings. Its occurrence increases with increasing latitude and continentality. An oxylophytic species characteristic of Mor humus forms.

Equisetum telmateia
Giant horsetail

Equisetaceae
(Horsetail family)

A shade-tolerant/intolerant, submontane to montane, European and Western North American horsetail (mainly in the Pacific region, less in the Cordilleran region, marginal in the Central region). Occurs in maritime to submaritime cool mesothermal climates on fresh to very moist, nitrogen-rich soils (Moder or Mull humus forms). Inhabits exposed mineral soil in broad-leaved stands on water-receiving (floodplain, seepage, and stream-edge) sites with fast-flowing groundwater near the ground surface. Its occurrence decreases with increasing latitude, elevation, and continentality. A nitrophytic species characteristic of alluvial floodplain forests.

Erigeron peregrinus
Subalpine daisy

Asteraceae
(Aster family)

A very shade-intolerant, subalpine to alpine, Asian and Western North American forb distributed equally in the Pacific and Cordilleran regions. Occurs on alpine tundra and subalpine boreal climates on moist to wet, nitrogen-rich soils (Moder and Mull humus forms). Common in high-elevation meadows, less often in subalpine parkland on water-receiving sites with friable organic materials. Its occurrence increases with increasing latitude. Characteristic of alpine and subalpine communities.

Eriophorum angustifolium
Eriophorum polystachyon
Narrow-leaved cotton-grass

Cyperaceae
(Sedge family)

A very shade-intolerant, submontane to sub-alpine, circumpolar sedge (transcontinental in North America). Occurs on wet to very wet, nitrogen-poor soils (Mor humus forms) within boreal, cool temperate, and cool mesothermal climates. Common in semi-terrestrial communities on water-collecting sites (peat bogs). Associated with *Kalmia occidentalis*, *Ledum groenlandicum*, *Sphagnum* species, and *Trientalis arctica*. An oxylophytic species characteristic of nutrient-poor wetlands.

Eriophyllum lanatum
Woolly sunflower, woolly eriophyllum

Asteraceae
(Aster family)

A very shade-intolerant, sub-montane to montane, Western North American forb distributed equally in the Pacific and Cordilleran regions. Occurs in cool temperate and cool meso-thermal climates on excessively dry to very dry, nitrogen-medium soils. Its occurrence increases with increasing precipitation and elevation. Grows in non-forested, grassy communities very shallow, on water-shedding sites; often inhabits disturbed sites. Characteristic of moisture-deficient sites.

Erythronium oregonum
White fawn lily

Liliaceae
(Lily family)

A shade-intolerant, submontane to montane, Pacific North American forb. Occurs in maritime to submaritime summer-dry cool mesothermal climates on moderately dry to fresh, nitrogen-medium soils.-Common in meadow-like communities and open-canopy forests on water-shedding sites. Its occurrence decreases with increasing elevation and continentality. Characteristic of dry mesothermal forests.

Erythronium revolutum
Pink fawn lily

Liliaceae
(Lily family)

A shade-tolerant/intolerant, submontane to montane, Pacific North American forb. Occurs in maritime to submaritime summer-wet cool mesothermal climates on fresh to moist, nitrogen-rich soils (Moder and Mull humus forms). Sporadic in broad-leaved forests on water-receiving sites; its occurrence decreases with increasing latitude, elevation, and continentality. Characteristic of alluvial floodplain forests.

Fauria crista-galli
Nephrophyllidium crista-galli
Deer-cabbage

Menyanthaceae
(Buckbean family)

A very shade-intolerant, sub-montane to subalpine, Pacific North American forb; its occurrence decreases with increasing continentality. Occurs in hypermaritime to maritime subalpine boreal and wet cool mesothermal climates on wet to very wet, nitrogen-poor soils (Mor humus forms). Common in semi-terrestrial communities on water-collecting sites. Inhabits depressions with groundwater table at or above ground surface. Usually associated with *Sphagnum* species. An oxylophytic species characteristic of nutrient-poor wetlands.

Festuca occidentalis
Western fescue

Poaceae
(Grass family)

A shade-tolerant/intolerant, submontane to montane, North American grass distributed in the Pacific, Cordilleran, and Central regions. Occurs on nitrogen-poor soils within boreal, cool temperate, and cool mesothermal climates. Its occurrence increases with increasing continentality, and decreases with increasing elevation and latitude. Scattered in coniferous forests on shallow and/or strongly drained soils on stony and rocky sites. An oxylophytic species characteristic of Mor humus forms.

Festuca subulata
Bearded fescue
Festuca subuliflora
Crinkle-awned fescue

Poaceae
(Grass family)

Shade-tolerant/intolerant, submontane to montane, Pacific North American grasses; their occurrence decreases with increasing elevation and continentality. These species occur in maritime to submaritime cool mesothermal climates on fresh to very moist, nitrogen-rich soils. Scattered in coniferous forests, more frequent in broad-leaved forests on water-receiving (floodplain and seepage) sites. Frequently associated with *Galium triflorum*, *Polystichum munitum*, and *Tiarella trifoliata*. Nitrophytic species characteristic of Moder and Mull humus forms.

Fragaria vesca
Fragaria bracteata
Wood strawberry

Rosaceae
(Rose family)

A shade-tolerant/intolerant, submontane to montane, circumpolar forb (transcontinental in North America). Occurs on moderately dry to fresh, nitrogen-medium soils within boreal, wet temperate, and cool mesothermal climates; its occurrence increases with increasing continentality, and decreases with increasing elevation. Inhabits exposed mineral soils on water-shedding sites; common in early-seral, meadow-like communities; sporadic in open-canopy, young-seral forests. Characteristic of disturbed sites.

Fragaria virginiana
Wild strawberry

Rosaceae
(Rose family)

A shade-intolerant, submontane to subalpine, circumpolar forb (transcontinental in North America). Occurs on nitrogen-medium soils within boreal, temperate, cool semiarid, and cool mesothermal climates. Its occurrence increases with increasing continentality, and decreases with increasing elevation. Frequently inhabits exposed, calcium-rich, mineral soil water-shedding sites. Common in non-forested (grassy) communities, less frequent in open-canopy forests. Characteristic of young-seral forests.

Fritillaria lanceolata
Chocolate lily

Liliaceae
(Lily family)

A shade-intolerant, submontane to montane, Western North American forb distributed more in the Pacific than the Cordilleran region. Occurs in maritime to submaritime cool mesothermal climates on nitrogen-rich soils. Its occurrence decreases with increasing precipitation, elevation, and continentality. Sporadic in non-forested grassy or meadow-like communities (less frequent in open-canopy forests) on water-shedding and water-receiving sites. Characteristic of Moder and Mull humus forms.

Galium aparine
Cleavers

Rubiaceae
(Madder family)

A shade-tolerant/intolerant, submontane to montane, circumpolar forb (transcontinental in North America). Occurs in cool temperate and cool mesothermal climates on nitrogen-rich soils; its occurrence decreases with increasing elevation. Common in herbaceous communities on disturbed sites; sporadic in open-canopy, broad-leaved forests on water-shedding and water-receiving sites. A nitrophytic species characteristic of Moder and Mull humus forms.

Galium triflorum
Sweet-scented bedstraw

Rubiaceae
(Madder family)

A shade-tolerant, submontane to subalpine, circumpolar forb (transcontinental in North America). Occurs on fresh to very moist, nitrogen-rich soils within boreal, temperate, and cool mesothermal climates; its occurrence decreases with increasing elevation. Common on water-receiving (alluvial, floodplain, seepage, and stream-edge) sites; also present in early-seral communities. Scattered in coniferous forests, plentiful in broad-leaved forests. Typically associated with *Athyrium filix-femina*, *Polystichum munitum*, *Rubus parviflorus*, and *Tiarella trifoliata*. A nitrophytic species characteristic of Moder and Mull humus forms.

Gaultheria humifusa
Alpine-wintergreen

Ericaceae
(Heath family)

A shade-intolerant, subalpine to alpine, Western North American evergreen shrub distributed equally in the Pacific and Cordilleran regions. Occurs in alpine tundra and subalpine boreal climates on fresh to very moist, nitrogen-poor soils. Common in non-forested (heath and krummholz) and open-canopy forested communities on water-shedding sites with prolonged snow-lie. Often grows with *Rhododendron albiflorum* and *Vaccinium membranaceum*. An oxylophytic species characteristic of Mor humus forms.

Gaultheria ovatifolia
Western tea-berry

Ericaceae
(Heath family)

A shade-tolerant/intolerant, montane to subalpine, Western North American evergreen shrub distributed less in the Pacific than the Cordilleran region. Occurs predominantly in continental subalpine boreal and cool temperate climates on very dry to moderately dry, nitrogen-poor soils. Scattered in coniferous forests on water-shedding sites on eastern Vancouver Island and in the coast-interior ecotone; its occurrence increases with increasing continentality. Commonly associated with *Pleurozium schreberi, Rhytidiopsis robusta*, and *Vaccinium membranaceum*. An oxylophytic species characteristic of Mor humus forms.

Gaultheria shallon
Salal

Ericaceae
(Heath family)

A shade-tolerant/intolerant, submontane to montane, Western North American evergreen shrub distributed mainly in the Pacific and marginally in the Cordilleran region. Occurs in hypermaritime to maritime cool mesothermal climates on nitrogen-poor soils; its occurrence decreases with increasing elevation and continentality. Often dominant in open-canopy coniferous forests on water-shedding sites; forms thickets on cutover areas with relatively undisturbed forest floors. On nutrient-rich sites, restricted to decaying coniferous wood; on wet sites, on topographic prominences. Absent or sporadic in the shaded understory of immature, closed-canopy stands. Due to its extensive horizontal root system in the uppermost soil layer and decay-resistant foliage and roots, high cover of salal reduces available soil water and/or decomposition of forest floor materials. These features hinder forest regeneration and growth, particularly on moisture-deficient sites. An oxylophytic species characteristic of Mor humus forms.

Gentiana douglasiana
Swamp gentian
Gentiana sceptrum
King gentian

Gentianaceae
(Gentian family)

Very shade-intolerant, submontane to montane, Pacific North American forbs; their occurrence decreases with increasing continentality. Both species occur in hypermaritime to submaritime cool mesothermal climates on wet to very wet, nitrogen-poor (*G. douglasiana*) to nitrogen-medium (*G. sceptrum*) soils. Occasional in semi-terrestrial communities (less frequently in open-canopy forests) on water-collecting sites with surface groundwater table and accumulations of peat. Characteristic of wetlands.

Geocaulon lividum
Comandra livida
Bastard toad-flax

Santalaceae
(Sandalwood family)

A shade-tolerant/intolerant, montane to subalpine, transcontinental North American forb (rare in the Pacific region). Occurs in continental boreal and cool temperate climates on nitrogen-poor soils. Scattered in coniferous forests on water-shedding and water-receiving sites in the central and northern coast-interior ecotone; its occurrence increases with increasing continentality. Often associated with *Hylocomium splendens*, *Pleurozium schreberi*, and *Ptilium crista-castrensis*. An oxylophytic species characteristic of Mor humus forms.

Geranium molle
Dovefoot geranium

Geraniaceae
(Geranium family)

A shade-intolerant, submontane to montane, Eurasian forb introduced to Pacific and Atlantic North America. Occurs in maritime to submaritime cool mesothermal climates on very dry to moderately dry, nitrogen-medium soils. Occasional in early-seral communities, sporadic in open-canopy forests on very shallow, water-shedding sites following disturbance. Its occurrence decreases with increasing elevation and continentality. Characteristic of disturbed sites.

Geum macrophyllum
Large-leaved avens

Rosaceae
(Rose family)

A shade-intolerant, submontane to subalpine, Asian and North American forb distributed in the Pacific, Cordilleran, and Atlantic regions. Occurs on fresh to very moist, nitrogen-rich soils (Moder and Mull humus forms), often with a fluctuating groundwater table. Widespread in subalpine boreal, temperate, and cool mesothermal climates; its occurrence decreases with increasing elevation. Usually inhabits exposed mineral soil; sporadic in open-canopy, broad-leaved forests on water-receiving (floodplain, seepage, and stream-edge) sites; scattered in early-seral communities. A nitrophytic species characteristic of disturbed sites.

Goodyera oblongifolia
Rattlesnake-plantain

Orchidaceae
(Orchid family)

A shade-tolerant, submontane to subalpine, transcontinental North American forb. Occurs on moderately dry to fresh, nitrogen-poor soils within boreal, temperate, and cool mesothermal climates. Common in coniferous forests on water-shedding sites; on nutrient-rich sites it inhabits decaying wood. Usually associated with *Hylocomium splendens*, *Pleurozium schreberi*, *Rhytidiadelphus loreus*, and *Rhytidiopsis robusta*. An oxylophytic species characteristic of Mor humus forms.

Gymnocarpium dryopteris
Oak fern

Aspleniaceae
(Spleenwort family)

A shade-tolerant, submontane to subalpine, circumpolar fern (transcontinental in North America). Occurs on fresh to very moist, nitrogen-rich soils within boreal, cool temperate, and cool mesothermal climates. Its occurrence increases with increasing precipitation, latitude, and continentality, and decreases with increasing temperature. Scattered to plentiful in forests on water-receiving (floodplain, seepage, and stream-edge) sites. Often associated with *Athyrium filix-femina, Oplopanax horridus, Streptopus roseus,* and *Tiarella unifoliata.* Characteristic of Moder and Mull humus forms.

Hemitomes congestum
Gnome-plant

Monotropaceae
(Indian pipe family)

A shade-tolerant/intolerant, submontane to montane, Western North American parasite distributed more in the Pacific than the Cordilleran region. Occurs in cool temperate and cool mesothermal climates on fresh to very moist, nitrogen-medium soils; its occurrence decreases with increasing elevation and continentality. Sporadic in the mossy understory of coniferous forests on water-shedding sites. Characteristic of moist coniferous forests.

Heracleum lanatum
Heracleum spondylium ssp. *montanum*
Cow-parsnip

Apiaceae
(Parsley family)

A shade-tolerant/intolerant, submontane to subalpine, Asian and transcontinental North American forb. Occurs on fresh to very moist, nitrogen-rich soils (often with a fluctuating groundwater table) within boreal, cool temperate, and cool mesothermal climates; its occurrence increases with increasing continentality. Scattered in open-canopy forests (often abundant in early seral communities) on water-receiving (floodplain and seepage) sites. A nitrophytic species characteristic of Moder and Mull humus forms.

Herbertus aduncus

Herbertaceae

A shade-tolerant/intolerant, submontane to montane, circumpolar liverwort distributed in Pacific and Atlantic North America. Occurs in wet cool mesothermal climates on stems and branches of old growth conifers (usually Sitka spruce), occasionally on forest floor in hypermaritime climates, less often on coarse fragments. Its occurrence decreases with increasing elevation and continentality. Characteristic of wet mesothermal forests.

Heuchera micrantha
Small-flowered alumroot

Saxifragaceae
(Saxifrage family)

A shade-tolerant/intolerant, submontane to subalpine, Western North American forb distributed more in the Pacific than the Cordilleran region. Occurs in maritime to submaritime cool mesothermal climates on nitrogen-rich soils; its occurrence decreases with increasing elevation and continentality. Scattered in non-forested communities and open-canopy forests on water-shedding (colluvial) and water-receiving (seepage and stream-edge) sites. Characteristic of Moder and Mull humus forms.

Hieracium albiflorum
White-flowered hawkweed

Asteraceae
(Aster family)

A shade-tolerant/intolerant, submontane to subalpine, Western North American forb distributed in the Pacific and the Cordilleran regions, marginally in the Central region. Occurs on moderately dry to fresh within boreal, cool temperate, and mesothermal climates. Commonly inhabits exposed mineral soil; plentiful on water-shedding sites in early-seral communities, less often in open-canopy forests. Characteristic of disturbed sites.

Hippuris montana
Mountain mare's-tail

Hippuridaceae
(Mare's-tail family)

A shade-intolerant, subalpine, Asian and Western North American forb distributed more in the Pacific than the Cordilleran region. Occurs in alpine tundra and subalpine boreal climates on very moist to wet, nitrogen-medium soils. Its occurrence increases with increasing latitude, and decreases with increasing continentality. Common in high-elevation, heath-like or low-shrub communities on water-shedding and water-receiving sites. Associated with *Cassiope mertensiana, Luetkea pectinata,* and *Phyllodoce empetriformis.* Characteristic of subalpine communities.

Holcus lanatus
Yorkshire fog

Poaceae
(Grass family)

A shade-intolerant, submontane to montane, European grass introduced to North America (transcontinental in North America). Occurs on fresh to very moist, nitrogen-medium soils within temperate, cool semiarid, and mesothermal climates; its occurrence decreases with increasing elevation. Commonly inhabits exposed mineral soil; scattered to plentiful in early-seral communities on water-shedding and water-receiving sites. Characteristic of disturbed sites.

Holodiscus discolor
Ocean-spray

Rosaceae
(Rose family)

A shade-tolerant/intolerant, submontane to montane, Western North American deciduous shrub distributed equally in the Pacific and Cordilleran regions. Occurs in cool temperate and cool mesothermal climates on very dry to moderately dry, nitrogen-medium soils; its occurrence decreases with increasing elevation, precipitation, and latitude. Scattered to plentiful in open and open-canopy, seral (usually Douglas-fir) forests on disturbed, water-shedding sites. Often associated with *Mahonia nervosa* and *Kindbergia oregana*. Characteristic of moisture-deficient sites.

Homalothecium megaptilum
Trachybryum megaptillum

Brachytheciaceae

A shade-tolerant, montane to subalpine, Western North American moss distributed more in the Pacific than the Cordilleran region. Occurs in maritime to submaritime cool mesothermal climates on very dry to moderately dry, nitrogen-poor soils. Sporadic in Douglas-fir forests on water-shedding sites; its occurrence decreases with increasing precipitation. Typically associated with *Gaultheria shallon, Hylocomium splendens* and *Kindbergia oregana*. An oxylophytic species characteristic of Mor humus forms.

Hookeria lucens
Hookeria acutifolia

Hookeriaceae

Very shade-tolerant, submontane to montane, circumpolar liverworts distributed more in Pacific than Cordilleran North America. Both species occur in hypermaritime to maritime summer-wet cool mesothermal climates on very moist to wet, nitrogen-poor soils; their occurrence increases with increasing precipitation, and decreases with increasing elevation.

Occasional in coniferous forests on water-receiving (seepage and stream-edge) sites, often on decaying coniferous wood. Oxylophytic species characteristic of Mor humus forms.

Huperzia selago
Lycopodium selago
Fir club-moss

Lycopodiaceae
(Club-moss family)

A shade-intolerant, submontane to alpine, circumpolar club-moss (transcontinental in North America). Occurs on moderately dry to fresh, nitrogen-poor soils within tundra, boreal, cool temperate, and cool mesothermal climates; its occurrence increases with increasing elevation. Frequent in montane and subalpine coniferous forests on water-shedding sites, more frequent in alpine communities. An oxylophytic species characteristic of Mor humus forms.

Hylocomium splendens

Hylocomiaceae

A shade-tolerant, submontane to subalpine, circumpolar moss (transcontinental in North America). Occurs on nitrogen-poor soils within tundra, boreal, temperate, and mesothermal climates; its occurrence decreases with increasing elevation and temperature. Abundant and often dominant in coniferous forests on water-shedding and water-receiving sites; on nutrient-rich sites inhabits decaying coniferous wood. Typically associated with *Gaultheria shallon, Paxistima myrsinites, Pleurozium schreberi, Rhytidiadelphus loreus, Vaccinium membranaceum,* and *V. parvifolium.* An oxylophytic species characteristic of Mor humus forms.

Hypericum formosum
Western St. John's-wort

Clusiaceae
(St. John's-wort family)

A shade-intolerant, submontane to alpine, Western North American forb distributed in the Pacific and Cordilleran regions. Occurs on fresh to very moist, nitrogen-medium soils within subalpine boreal, temperate, and mesothermal climates; its occurrence decreases with increasing temperature and elevation. Scattered in early-seral communities, sporadic in open-canopy coniferous forests on water-shedding and water-receiving sites. Characteristic of disturbed sites.

Hypochaeris radicata
Hairy cat's-ear

Asteraceae
(Aster family)

A very shade-intolerant, sub-montane to montane, European forb introduced to Pacific, Cordilleran, and Atlantic North America. Inhabits exposed mineral soil on water-shedding sites within montane boreal, temperate, and cool mesothermal climates; its occurrence decreases with elevation and latitude. Common in early-seral (often meadow-like) communities. Characteristic of disturbed sites.

Hypopithys lanuginosa
Hypopithys monotropa
Monotropa hypopithys
Pinesap

Monotropaceae
(Indian-pipe family)

A shade-tolerant/intolerant, submontane to subalpine, transcontinental North American parasite. Occurs on moderately dry to fresh, nitrogen-poor soils within boreal, cool temperate, and cool mesothermal climates; its occurrence decreases with elevation. Sporadic in coniferous forests on water-shedding sites. An oxylophytic species characteristic of Mor humus forms.

Isopterygium elegans   Hypnaceae

A very shade-tolerant, submontane to subalpine, European and North American moss distributed in the Pacific, Cordilleran and Atlantic regions. Occurs on fresh to very moist, nitrogen-poor soils within subalpine boreal, cool temperate, and cool mesothermal climates; its occurrence increases with increasing precipitation. Common in coniferous forests on exposed eluviated mineral soil or base-poor coarse fragments and bedrock on strongly drained, water-shedding sites. Often grows with *Rhytidiadelphus loreus* and *Plagiothecium undulatum.* An oxylophytic species characteristic of acid substrates.

Isothecium stoloniferum   Brachytheciaceae

A very shade-tolerant, submontane to subalpine, North American moss distributed in the Pacific, Cordilleran (uncommon), and Atlantic regions. Occurs in hypermaritime to maritime cool mesothermal climates on water-receiving sites; its occurrence decreases with increasing elevation and continentality. On moist and wet sites it grows on stems and branches, on fresh sites it inhabits coarse fragments and bedrock. Characteristic of mesothermal forests.

Juncus ensifolius
Dagger-leaved rush
Juncus effusus
Common rush

Juncaceae
(Rush family)

J. effusus - a very shade-intolerant, submontane to subalpine, circumpolar rush (trans-continental in North America). *J. ensifolius* - a very shade-intolerant, submontane to subalpine, Asian and Western North American rush (uncommon in the Central region).

These species occur on very moist to wet, nitrogen-medium soils within subalpine boreal, temperate, and mesothermal climates. Common and often dominant in early-seral communities on water-receiving and water-collecting sites with exposed and compacted mineral soil and a fluctuating groundwater table. Frequently associated with *Carex* species, *Deschampsia caespitosa* and *Scirpus microcarpus*. Characteristic of waterlogged sites.

Juniperus scopulorum
Rocky Mountain juniper

Cupressaceae
(Cypress family)

A very shade-intolerant, montane, Western North American evergreen coniferous shrub distributed more in the Cordilleran than the Pacific region. Occurs predominantly in continental cool temperate and cool semiarid climates on excessively dry to very dry and nitrogen-medium (often alkaline) soils; its occurrence increases with increasing continentality and temperature. In the coastal region, very sporadic in open-canopy shrub communities on very shallow, water-shedding sites of calcium-rich rock outcrops; common in the coast-interior ecotone. Characteristic of moisture-deficient sites.

Juniperus sibirica
Juniperus communnis
Common juniper

Cupressaceae
(Cypress family)

A shade-intolerant, submontane to alpine, circumpolar evergreen coniferous shrub (transcontinental in North America). Occurs in tundra, boreal, wet cool temperate, and cool mesothermal climates on very dry to moderately dry, nitrogen-medium soils; its occurrence increases with increasing latitude and continentality. Very sporadic in early-seral communities on water-shedding sites. Characteristic of disturbed sites.

Kalmia occidentalis
Kalmia polifolia
Kalmia microphylla ssp. *occidentalis*
Bog-laurel

Ericaceae
(Heath family)

A very shade-intolerant, montane to alpine, Western North American evergreen shrub distributed more in the Cordilleran than the Pacific region. Occurs predominantly in tundra and boreal climates on wet to very wet, nitrogen-poor soils (Mor humus forms); its occurrence increases with increasing latitude and continentality. Scattered in non-forested, semi-terrestrial communities on water-collecting sites with accumulations of peat. Commonly associated with *Ledum groenlandicum* and *Sphagnum* species. An oxylophytic species characteristic of nutrient-poor wetlands.

Kindbergia oregana
Stokesiella oregana
Eurhynchium oreganum

Brachytheciaceae

A shade-tolerant, submontane to subalpine, Western North American moss distributed more in the Pacific than the Cordilleran region. Occurs in cool mesothermal climates on moderately dry to fresh soils; its occurrence decreases with increasing elevation, precipitation, and continentality. Plentiful and often dominant in the understory of young-seral coniferous forests on water-shedding sites. Occasionally inhabits stems and branches, partly decomposed wood, and coarse fragments. Commonly associated with *Achlys triphylla*, *Gaultheria shallon*, *Hylocomium splendens*, *Mahonia nervosa*, or *Rhytidiadelphus loreus*. Characteristic of submontane mesothermal forests.

Kindbergia praelonga
Stokesiella praelonga
Eurhynchium praelongum

Brachytheciaceae

A very shade-tolerant, submontane to montane, circumpolar moss distributed in Pacific, Cordilleran, and Atlantic North America. Occurs on very moist to wet, nitrogen-rich soils often with fluctuating groundwater table. Widespread in boreal, cool temperate, and cool mesothermal climates; its occurrence decreases with increasing elevation. Common to plentiful on thin and friable forest floors in shaded forest understories on water-receiving and water-collecting sites. Typically associated with *Athyrium filix-femina*, *Lysichitum americanum*, *Oplopanax horridus*, *Pellia neesiana*, *Plagiomnium insigne*, and *Rubus spectabilis*. Characteristic of Moder and Mull humus forms.

Lathyrus nevadensis
Purple peavine

Fabaceae
(Pea family)

A shade-tolerant/intolerant, submontane to montane, Western North American forb distributed in the Pacific and Cordilleran regions. Occurs on moderately dry to fresh, nitrogen-rich soils within boreal, temperate, and cool mesothermal climates; its occurrence decreases with increasing elevation, precipitation, and latitude. Occasional in coniferous forests, frequent in broad-leaved forests on water-shedding sites. Symbiotic relationship with nitrogen-fixing bacteria enhances the supply of available soil nitrogen. Characteristic of Moder and Mull humus forms.

Lathyrus ochroleucus
Creamy peavine

Fabaceae
(Pea family)

A shade-tolerant/intolerant, montane, North American forb distributed in the Cordilleran and Central regions. Occurs in continental boreal and wet cool temperate climates on moderately dry to fresh, nitrogen-rich soils in the coast-interior ecotone; its occurrence increases with increasing continentality and latitude, and decreases with increasing elevation. Common in semi-open forests on base-rich, water-shedding and water-receiving sites. Symbiotic relationship with nitrogen-fixing bacteria enhances the supply of available soil nitrogen. Characteristic of Moder and Mull humus forms.

Ledum groenlandicum
Labrador tea

Ericaceae
(Heath family)

A shade-intolerant, submontane to subalpine, transcontinental North American evergreen shrub. Occurs on wet to very wet, nitrogen-poor soils (Mor humus forms) within boreal, cool temperate, and cool mesothermal climates; its occurrence decreases with increasing temperature. Common in non-forested or open-canopy, semi-terrestrial, *Sphagnum*-dominated communities on water-collecting sites with a stagnant water table and peat accumulations. An oxylophytic species characteristic of nutrient-poor wetlands.

Lepidozia reptans

Lepidoziaceae

A shade-tolerant, submontane to subalpine, circumpolar liverwort (transcontinental in North America). Occurs on fresh to very moist, nutrient-poor soils with boreal, cool temperate, and cool mesothermal climates. Plentiful in coniferous forests on water-shedding and water-receiving sites; usually inhabits decaying wood. Sporadic on the forest floor. Usually associated with *Calypogeia trichomanis*, *Rhizomnium glabrescens*, and *Scapania bolanderi*. An oxylophytic species characteristic of Mor humus forms.

Leptarrhena pyrolifolia
Leatherleaf saxifrage

Saxifragaceae
(Saxifrage family)

A shade-intolerant, subalpine, Western North American forb distributed equally in the Pacific and Cordilleran regions. Occurs in alpine tundra and subalpine boreal climates on very moist to wet, nitrogen-medium soils. Common in non-forested, semi-terrestrial (stream-edge, spring, and intermittent stream), and meadow-like communities on water-receiving (flooded) sites. Usually associated with *Caltha leptosepala*, *Petasites frigidus*, and *Philonotis fontana*. Characteristic of alpine and subalpine communities.

Letharia vulpina

Usneaceae

A shade-intolerant, montane to subalpine, circumpolar epiphytic lichen (transcontinental in North America; less frequent in the Pacific region, marginal in the Central region). Occurs in cool temperate and cool semiarid climates on the coast-interior ecotone; its occurrence increases with increasing continentality, and decreases with increasing elevation. Scattered to plentiful on the bark of branches and stems of mature, often isolated, coniferous trees (a corticolous species) on water-shedding sites. Characteristic of dry and warm continental forests.

Leucolepis menziesii
Leucolepis acanthoneura

Mniaceae

A very shade-tolerant, submontane to montane, Western North American moss distributed more in the Pacific than the Cordilleran region. Occurs in cool mesothermal climates on very moist to wet, nitrogen-rich soils; its occurrence decreases with increasing elevation and continentality. Common on water-receiving (seepage and flooded) sites. Inhabits friable mineral soils rich in organic matter. Typically associated with *Kindbergia praelonga*, *Polystichum munitum*, *Plagiomnium insigne*, and *Tiarella trifoliata*. Characteristic of Moder and Mull humus forms.

Lilium columbianum
Tiger lily

Liliaceae
(Lily family)

A shade-intolerant, submontane to subalpine, Western North American forb distributed equally in the Pacific and Cordilleran regions. Occurs on moderately dry to fresh, nitrogen-medium soils within boreal, cool temperate, and cool mesothermal climates. Scattered in early-seral communities, occasional in forested communities on water-shedding, often on colluvial, sites. Characteristic of open-canopy, young-seral forests.

Linnaea borealis
Twinflower

Caprifoliadeae
(Honeysuckle family)

A shade-tolerant/intolerant, submontane to subalpine, circumpolar evergreen shrub (transcontinental in North America). Occurs predominantly on moderately dry to fresh soils within boreal, cool temperate, and cool mesothermal climates. Scattered to plentiful in mossy understories in coniferous forests on water-shedding sites; on moist and wet sites it inhabits topographic prominences, on nutrient-rich sites it is restricted to decaying coniferous wood; persists on undisturbed cutover sites. Commonly associated with

Cornus canadensis, Goodyera oblongifolia, Hylocomium splendens, Kinbergia oregana, and *Vaccium parvifolium.* Characteristic of Moder and acidic Moder humus forms.

Listera caurina
Northwestern twayblade

Orchidaceae
(Orchid family)

A shade-tolerant, submontane to montane, Western North American forb distributed more in the Pacific than the Cordilleran region. Occurs on fresh to very moist, nutrient-medium soils within boreal, wet cool temperate, and cool mesothermal climates; its occurrence decreases with increasing elevation and continentality. Scattered in dense coniferous forests on water-shedding and water-receiving sites; often associated with *Blechnum spicant, Gaultheria shallon, Hylocomium splendens, Rhytidiadelphus loreus, Plagiothecium undulatum,* and *Vaccinium alaskaense.* Characteristic of friable Mors and acidic Moder humus forms.

Listera convallarioides
Broad-leaved twayblade

Orchidaceae
(Orchid family)

A shade-tolerant, submontane to montane, Asian and transcontinental North American forb. Occurs in cool temperate and cool mesothermal climates on fresh to very moist, nitrogen-rich soils; its occurrence decreases with increasing elevation and continentality. Sporadic in semi-terrestrial communities on water-collecting sites; occasionally associated with *Lysichitum americanum*. A nitrophytic species characteristic of Moder and Mull humus forms.

Listera cordata
Heart-leaved twayblade

Orchidaceae
(Orchid family)

A shade-tolerant/intolerant, montane to subalpine, circumpolar forb (transcontinental in North America). Occurs on moderately dry to fresh, nitrogen-poor soils within boreal, cool temperate, and cool mesothermal climates. Common in coniferous forests on water-shedding and water-receiving sites; often grows with *Listera caurina*, *Gaultheria shallon*, *Hylocomium splendens*, *Rhytidiadelphus loreus*, and *Vaccinium parvifolium*. An oxylophytic species characteristic of Mor humus forms.

Lobaria oregana   *Stictaceae*

A shade-tolerant/intolerant, submontane to montane, Western North American lichen distributed more in the Pacific than the Cordilleran region. Contains nitrogen-fixing blue-green algae. Scattered to plentiful on branches and stems of dominant trees in semi-open, mature forests on a variety of sites in cool mesothermal climates. Commonly found on the ground after windstorms. Characteristic of mesothermal forests.

Loiseleuria procumbens   *Ericaceae*
Alpine-azalea   (Heath family)

A very shade-intolerant, montane to alpine, circumpolar evergreen shrub (transcontinental in North America). Occurs on very moist to wet, nitrogen-poor soils (Mor humus forms), within alpine tundra, boreal, cool temperate, and cool mesothermal climates; its occurrence increases with increasing latitude. Frequent in heath communities, occasionally found in parkland-like forests on water-shedding and near sea-level in hypermaritime muskegs of north coast. Associated with *Barbilophozia floerkei, Cassiope mertensiana, Phyllodoce empetriformis*, and *Vaccinium uliginosum*. An oxylophytic species characteristic of alpine and subalpine communities.

Lomatium dissectum
Fern-leaved desert-parsley

Apiaceae
(Parsley family)

A shade-tolerant/intolerant, montane to subalpine, Western North American forb distributed equally in the Pacific and Cordilleran regions, marginally in the Central region. Occurs on very dry to moderately dry, nitrogen-rich soils (Moder and Mull humus forms) within subalpine boreal, temperate, cool semiarid, and dry mesothermal climates. Sporadic in open-canopy forests on water-shedding sites. A nitrophytic species characteristic of moisture-deficient sites.

Lonicera hispidula
Purple honeysuckle, hairy honeysuckle
Lonicera ciliosa
Orange honeysuckle, western trumpet honeysuckle

Caprifoliaceae
(Honeysuckle family)

Shade-tolerant/intolerant, submontane to montane, Western North American shrubs (vines) distributed more in the Pacific than the Cordilleran regions. These species occur in cool temperate and cool mesothermal climates (*L. ciliosa*) and maritime to submaritime summer-dry mesothermal climates (*L. hispidula*) on very dry to moderately dry, nitrogen-medium soils. Sporadic on water-shedding sites; climb up shrubs and trees in open-canopy Douglas-fir forests. Characteristic of moisture-deficient sites.

Lonicera involucrata
Black twinberry, black honeysuckle

Caprifoliaceae
(Honeysuckle family)

A shade-tolerant/intolerant, submontane to subalpine, transcontinental North American deciduous shrub. Occurs on very moist to wet, nitrogen-rich soils (Moder or Mull humus forms). Tolerates fluctuating groundwater tables. Widespread in boreal, temperate, and cool mesothermal climates; scattered to plentiful in the open or in broad-leaved forests on water-receiving (alluvial, floodplain, seepage, and stream-edge) sites and on water-collecting (swamps and fens) sites. Persists on cutover sites where it may hinder natural regeneration and growth of shade-intolerant conifers. A nitrophytic species characteristic of alluvial floodplain forests.

Lonicera utahensis
Utah honeysuckle, red twinberry

Caprifoliaceae
(Honeysuckle family)

A shade-tolerant/intolerant, montane to subalpine, Western North American deciduous shrub distributed more in the Cordilleran than the Pacific region. Occurs predominantly in continental subalpine boreal and cool temperate climates on nitrogen-medium soils; its occurrence increases with increasing elevation and continentality. Sporadic to scattered in herbaceous understories on water-shedding and water-receiving sites in the coast-interior ecotone. Characteristic of subalpine continental forests.

Luetkea pectinata
Partridgefoot

Rosaceae
(Rose family)

A very shade-intolerant, subalpine to alpine, Western North American forb distributed equally in the Pacific and Cordilleran regions. Occurs in alpine tundra and subalpine boreal climates on fresh to very moist, nitrogen-medium soils. Inhabits organic or mineral substrates receiving seepage from late-melting snowbanks. Abundant in heath communities and parkland-like forests; often grows with *Cassiope mertensiana* and *Phyllodoce empetriformis*. Characteristic of alpine and subalpine communities.

Lupinus arcticus
Arctic lupine

Fabaceae
(Pea family)

A shade-intolerant, montane to alpine, North American forb distributed in the Cordilleran and Central regions. Occurs in continental alpine tundra and boreal climates on nitrogen-rich (Moder or Mull humus forms); its occurrence increases with increasing latitude and continentality. Scattered to plentiful in early-seral and meadow-like communities in the coast-interior ecotone; often inhabits exposed mineral soil. Symbiotic relationship with nitrogen-fixing organisms enhances the supply of available soil nitrogen. Characteristic of disturbed sites.

Lupinus nootkatensis
Nootka lupine

Fabaceae
(Pea family)

A shade-tolerant/intolerant, submontane to montane, Western North American forb distributed more in the Pacific than the Cordilleran region. Occurs on nitrogen-rich soils (Moder or Mull humus forms) with boreal and temperate, and cool mesothermal climates. Sporadic in meadow-like communities and open-canopy forests on water-receiving sites; more common in early-seral communities on cutover and burnt sites where it inhabits exposed mineral soil. Symbiotic relationship with nitrogen-fixing organisms enhances the supply of available soil nitrogen. Characteristic of disturbed sites.

Luzula multiflora
Many-flowered woodrush

Juncaceae
(Rush family)

A shade-intolerant, submontane to montane, circumpolar graminoid (transcontinental in North America). Occurs in cool temperate and cool mesothermal climates on excessively dry to very dry, nitrogen-poor soils (Mor humus forms). Sporadic in non-forested communities and open-canopy Douglas-fir forests on shallow, rapidly drained, stony and rocky, water-shedding sites; often inhabits exposed mineral soils. An oxylophytic species characteristic of moisture-deficient sites.

Luzula parviflora
Small-flowered woodrush

Juncaceae
(Rush family)

A shade-tolerant/intolerant, submontane to subalpine, circumpolar graminoid (transcontinental in North America). Occurs on fresh to very moist, nitrogen-medium soils within boreal, wet temperate, and cool mesothermal climates. Sporadic on water-shedding and water-receiving sites. Often associated with *Blechnum spicant*, *Dryopteris expansa*, *Gymnocarpium dryopteris*, and *Polystichum munitum*. Characteristic of friable Mor and acidic Moder humus forms.

Lycopodium annotinum
Stiff club-moss

Lycopodiaceae
(Club-moss family)

A shade-tolerant/intolerant, montane to subalpine, circumpolar club-moss (transcontinental in North America). Occurs in boreal and cool temperate climates on moderately dry to fresh, nitrogen-medium soils; its occurrence increases with increasing latitude. Sporadic in non-forested communities and in coniferous forests on water-shedding sites (less frequently on water-receiving sites). Characteristic of boreal coniferous forests.

Lycopodium clavatum
Running club-moss

Lycopodiaceae
(Club-moss family)

A shade-tolerant/intolerant, submontane to subalpine, circumpolar club-moss (transcontinental in North America). Occurs on moderately dry to fresh, nitrogen-poor soils. Widespread on water-shedding sites in boreal, cool temperate, and cool mesothermal climates; its occurrence increases with increasing latitude. Common in open-canopy coniferous forests; often inhabits decaying wood, persists on undisturbed cutover areas. Grows with *Hylocomium splendens, Pleurozium schreberi, Ptilium crista-castrensis,* and *Rhytidiadelphus loreus.* An oxylophytic species characteristic of Mor humus forms.

Lycopodium obscurum
Lycopodium dendroideum
Ground-pine
Lycopodium complanatum
Ground-cedar

Lycopodiacea
(Club-moss family)

Shade-tolerant/intolerant, submontane to montane, club-mosses. *L. obscurum* - Asian and transcontinental North American (rare in Pacific region); *L. complanatum* - circumpolar (transcontinental in North America).

These species occur in continental montane boreal and cool temperate climates on moderately dry to fresh (*L. complanatum*), fresh to very moist (*L. obscurum*), nitrogen-poor soils. Their occurrence increases with increasing continentality and latitude. Occur in coniferous forests in the coast-interior ecotone. Usually associated with *Hylocomium splendens, Pleurozium schreberi,* and *Ptilium crista-castrensis.* Oxylophytic species characteristic of Mor humus forms.

Lycopodium sitchense
Mountain club-moss
Lycopodium alpinum
Alpine club-moss

Lycopodiaceae
(Club-moss family)

Shade-intolerant, subalpine to alpine clubmosses, *L. sitchense* Asian and North American transcontinental, *L. alpinum* circumpolar and transcontinental North American. Occurs in alpine tundra and boreal climates on fresh to very moist, nitrogen-poor soils. Common in high-elevation, non-forested heath communities (with *Cassiope mertensiana, Luetkea pectinata*, and *Phyllodoce empetriformis*) on water-shedding sites; their occurrence increases with increasing elevation and latitude. Oxylophytic species characteristic of alpine and subalpine communities.

Lysichitum americanum
Skunk-cabbage

Araceae
(Arum family)

A shade-tolerant/intolerant, submontane to subalpine, Western North American herb distributed more in the Pacific than the Cordilleran region. Occurs on wet to very wet, nitrogen-rich soils (Moder and Mull humus forms) within subalpine boreal, cool temperate, and cool mesothermal climates. Often dominant in non-forested, semi-terrestrial communities and in understories of open-canopy stands of red alder, Sitka spruce, yellow-cedar, or western redcedar on gleysolic or organic soils. Usually associated with *Athyrium filix-femina, Kindbergia praelonga*, and *Pellia neesiana* on water-collecting sites with slow-moving groundwater close to the ground surface. A nitrophytic species characteristic of nutrient-rich wetlands.

Madia madioides
Woodland tarweed

Asteraceae
(Aster family)

A shade-intolerant, submontane to montane, Western North American forb distributed more in the Pacific than the Cordilleran region. Occurs in maritime to submaritime summer-dry cool mesothermal climates on very dry to moderately dry, nitrogen-medium soils; its occurrence decreases with increasing latitude, elevation, and precipitation. Most frequent on exposed mineral soil of recently cleared and burnt, water-shedding sites. Characteristic of disturbed, moisture-deficient sites.

Mahonia aquifolium
Berberis aquifolium
Tall Oregon-grape

Berberidaceae
(Barberry family)

A shade-tolerant/intolerant, submontane to montane, Western North American evergreen shrub distributed more in the Cordilleran than the Pacific region. Occurs predominantly in continental cool temperate and cool semiarid climates on very dry to moderately dry, nitrogen-medium soils; its occurrence increases with increasing summer drought and continentality. Sporadic in summer-dry mesothermal climates, common in open-canopy Douglas-fir forests in the coast-interior ecotone. Often associated with *Agropyron spicatum*, *Calamagrostis rubescens*, and *Rhytidiadelphus triquetrus*. Characteristic of moisture-deficient sites.

Mahonia nervosa
Berberis nervosa
Dull Oregon-grape

Berberidaceae
(Barberry family)

A shade-tolerant/intolerant, submontane to montane, Pacific North American evergreen shrub. Occurs in maritime to submaritime cool mesothermal climates on moderately dry to fresh, nitrogen-medium soils; its occurrence decreases with increasing precipitation, elevation, and continentality. Scattered to abundant, occasionally dominant, in the understory of semi-open forests (persists on cutover sites) on water-shedding sites. Inhabits coarse-skeletal soils. Commonly associated with *Gaultheria shallon*, *Kindbergia oregana*, and *Polystichum munitum*. Characteristic of mesothermal forests.

Maianthemum dilatatum
False lily-of-the-valley

Liliaceae
(Lily family)

A shade-tolerant and ocean spray-tolerant, submontane to subalpine, Western North American forb distributed more in the Pacific than the Cordilleran region. Occurs in cool mesothermal climates on very moist to wet, nitrogen-rich soils (Moder or Mull humus forms); its occurrence decreases with increasing elevation and continentality. Scattered to plentiful in coniferous and broad-leaved forests on water-receiving and water-collecting sites, commonly found on stream-edge sites, floodplains, and sites affected by ocean spray. Grows with *Blechnum spicant*, *Polystichum munitum*, *Tiarella trifoliata*, *Trautvetteria caroliniensis*, and *Lysichitum americanum*. Characteristic of alluvial floodplain forests.

Malus fusca
Pyrus fusca
Pacific crab apple

Rosaceae
(Rose family)

A shade-intolerant, submontane to montane, Pacific North American shrub or broad-leaved tree. Occurs in cool mesothermal climates on wet to very wet, nitrogen-rich soils (Moder and Mull humus forms); its occurrence decreases with increasing elevation and continentality. Scattered in open-canopy forests on water-collecting sites; often inhabits brackish-water marshes and sites affected by ocean spray; rare on water-shedding sites. Characteristic of nutrient-rich wetlands.

Marchantia polymorpha

Marchantiaceae

A shade-intolerant, submontane to subalpine, cosmopolitan liverwort; (transcontinental in North America). Occurs on very moist to wet soils within boreal, temperate, mesothermal, and tropical climates. Inhabits exposed mineral soil on water-receiving and water-collecting sites. Characteristic of recently cleared and burnt sites.

Melica subulata
Alaska oniongrass

Poaceae
(Grass family)

A shade-tolerant/intolerant, submontane to montane, South American and Western North American grass distributed more in the Pacific than the Cordilleran region. Occurs in cool temperate and cool mesothermal climates on fresh to very moist, nitrogen-rich soils; its occurrence decreases with increasing latitude, elevation, and continentality. Scattered in herbaceous understories on water-receiving sites with melanized, base-rich soils. A nitrophytic species characteristic of Moder and Mull humus forms.

Menyanthes trifoliata
Buckbean

Menyanthaceae
(Buckbean family)

A very shade-intolerant, submontane to subalpine, circumpolar forb; (transcontinental in North America). Occurs on wet to very wet, nitrogen-medium soils within boreal, wet temperate, and cool mesothermal climates. Often dominant in aquatic communities along freshwater lakes; occasional on water-collecting sites. Characteristic of wetlands.

Menziesia ferruginea
False azalea

Ericaceae
(Heath family)

A shade-tolerant/intolerant, submontane to subalpine, Western North American deciduous shrub distributed equally in the Pacific and Cordilleran regions, and marginally in the Central region. Occurs on moderately dry to fresh, nitrogen-poor soils within boreal, cool temperate, and cool mesothermal climates; its occurrence increases with increasing precipitation. Scattered to plentiful in coniferous forests on water-shedding sites; with acid forest floors. On nutrient-rich sites, restricted to decaying coniferous wood. Typically associated with *Hylocomium splendens, Rhytidiadelphus loreus, Rhytidiopsis robusta, Vaccinium* species. An oxylophytic species characteristic of Mor humus forms.

Mitella breweri
Brewer's mitrewort
M. ovalis
oval-leaved mitrewort
M. pentandra
five-stamened mitrewort

Saxifragaceae
(Saxifrage family)

Shade-tolerant Western North American forbs which grow on boreal, temperate, and cool mesothermal climates. *Mitella breweri* and *M. pentandra* are montane to subalpine and their occurrence increases with increasing elevation and continentality. *M. ovalis* is restricted to the Pacific region, submontane to montane.

These species occur on very moist to wet, nitrogen-rich soils. Sparse to occasional in semi-open stands on water-receiving and water-collecting sites. Usually associated with *Athyrium filix-femina, Gymnocarpium dryopteris, Oplopanax horridus, Ribes lacustre, Tiarella trifoliata,* and *T. unifoliata*. Nitrophytic species characteristic of Moder and Mull humus forms.

Mitella nuda
Common mitrewort

Saxifragaceae
(Saxifrage family)

A shade-tolerant/intolerant, montane to subalpine, Asian and transcontinental North American forb (absent in the Pacific region). Occurs in continental boreal and cool temperate climates on fresh to moist, nitrogen-medium soils; its occurrence increases with increasing elevation and continentality. Sporadic in semi-open coniferous forests on water-shedding and water-receiving sites in the eastern coast-interior ecotone. Characteristic of continental forests.

Mnium spinulosum

Mniaceae

A shade-tolerant/intolerant, submontane to subalpine, circumpolar moss (transcontinental in North America). Occurs predominantly in continental boreal and cool temperate climates on moderately dry to fresh, nitrogen-medium soils; its occurrence increases with increasing continentality. Scattered to plentiful in coniferous forests on water-shedding sites in the coast-interior ecotone; frequently inhabits decaying wood. Characteristic of continental forests.

Moehringia macrophylla
Arenaria macrophylla
Big-leaved sandwort

Caryophyllaceae
(Pink family)

A shade-tolerant/intolerant, submontane to montane, Asian and transcontinental North American forb. Occurs in cool temperate and cool mesothermal climates on moderately dry to fresh, nitrogen-rich soils; its occurrence increases with increasing summer temperature and decreases with increasing latitude. Scattered in semi-open Douglas-fir forests on water-shedding sites. A nitrophytic species characteristic of Moder and Mull humus forms.

Moneses uniflora
Single delight

Pyrolaceae
(Wintergreen family)

A shade-tolerant/intolerant, submontane to montane, circumpolar herb (transcontinental in North America). Occurs on fresh to moist, nitrogen-medium soils within boreal, wet temperate, and cool mesothermal climates. Its occurrence increases with increasing precipitation and decreases with increasing elevation. Occurs sporadically in coniferous forests (occasional on decaying wood) on water-shedding and water-receiving sites. Characteristic of mycorrhiza-rich Mor and acidic Moder humus forms.

Monotropa uniflora
Indian-pipe

Monotropaceae
(Indian-pipe family)

A shade-tolerant, submontane to montane, Asian and transcontinental North American saprophyte. Occurs on fresh to moist, nitrogen-medium soils in montane boreal, wet temperate, and cool mesothermal climates; its occurrence decreases with increasing latitude. Occurs sporadically in closed-canopy coniferous forests (most common on southern Vancouver Island and the Gulf Islands) on water-shedding and water-receiving sites. Characteristic of mycorrhiza-rich Mor and acidic Moder humus forms.

Montia parvifolia
Small-leaved montia

Portulacaceae
(Purslane family)

A shade intolerant, submontane to montane, Western North American forb distributed equally in the Pacific and Cordilleran regions. Occurs in cool temperate and cool mesothermal climates on nitrogen-medium soils; its occurrence decreases with increasing latitude and elevation. Scattered in open-canopy forests on water-shedding and water-receiving sites. Usually inhabits very shallow, friable organic materials accumulated on stony and rocky sites affected by temporary surface flow of water and organic materials. Characteristic of colluvial sites.

Mycelis muralis
Lactuca muralis
Wall-lettuce

Asteraceae
(Aster family)

A shade-tolerant/intolerant, submontane to montane, European forb introduced to Pacific, Cordilleran, and Atlantic North America. Occurs in maritime to submaritime cool mesothermal climates on fresh to moist, nitrogen-rich soils; its occurrence decreases with increasing elevation and continentality. Frequent in semi-open, seral forests on disturbed water-shedding and water-receiving sites; common, occasionally dominant, in early-seral communities on cutover and burnt site (often inhabits exposed mineral soil). Often grows with *Achlys triphylla, Epilobium angustifolium, Kindbergia oregana, Polystichum munitum,* and *Tiarella trifoliata.* A nitrophytic species characteristic of Moder and Mull humus forms.

Myrica gale
Sweet gale

Myricaceae
(Bayberry family)

A very shade-intolerant, submontane to subalpine, circumpolar deciduous shrub (transcontinental in North America). Occurs on wet to very wet, nitrogen-medium soils within boreal, wet cool temperate, and cool mesothermal climates; its occurrence increases with increasing latitude. Symbiotic with nitrogen-fixing organisms. Common and often dominant in non-forested, semi-terrestrial communities on water-collecting sites. Occasionally grows with *Carex obnupta, C. sitchensis,* and *Spiraea douglasii.* Characteristic of wetlands.

Nuphar polysepalum
Nuphar luteum ssp. *polysepalum*
Yellow waterlily

Nymphaeaceae
(Waterlily family)

A very shade-intolerant, submontane to subalpine, Western North American forb distributed equally in the Pacific
and Cordilleran regions, and
marginally in the Central region. Occurs on wet to very wet
soils within boreal, temperate,
and cool mesothermal climates.
Occasional in non-forested,
semi-terrestrial communities
on water-collecting sites; dominant in initial stages of primary succession on aquatic
sites along freshwater lakes.
Characteristic of shallow water wetlands.

Oemleria cerasiformis
Osmaronia cerasiformis
Indian-plum

Rosaceae
(Rose family)

A shade-tolerant/intolerant, submontane to
montane, Western North American deciduous
shrub distributed more in the Pacific than the
Cordilleran region. Occurs in maritime to
submaritime summer-dry cool mesothermal
climates on fresh to very moist, nitrogen-rich
soils often with a fluctuating groundwater
table. Its occurrence decreases with increasing latitude and continentality. Scattered in
broad-leaved forests on water-receiving sites
(most often on floodplains) with melanized
and often gleyed soils. Commonly associated
with *Cornus sericea*, *Sambucus racemosa*, and
Symphoricarpos albus. Characteristic of Moder
and Mull humus forms.

Oenanthe sarmentosa
Pacific oenanthe, water-parsley

Apiaceae
(Parsley family)

A shade-tolerant/intolerant, submontane to montane, Western North American herb distributed more in the Pacific than the Cordilleran region. Occurs in cool mesothermal climates on wet to very wet, nitrogen-rich soils (Moder and Mull humus forms); its occurrence decreases with increasing latitude, elevation, and continentality. Occasional in forest understories on water-receiving sites with fluctuating groundwater table; common in non-forested, semi-terrestrial communities on water-collecting sites. Usually associated with *Lysichitum americanum*. A nitrophytic species characteristic of nutrient-rich wetlands.

Oplopanax horridus
Devil's club

Araliaceae
(Ginseng family)

A shade-tolerant, submontane to subalpine, Asian and North American deciduous shrub distributed equally in the Pacific, Cordilleran, and Central regions. Occurs on very moist to wet, nitrogen-rich soils within boreal, cool temperate, and cool mesothermal climates; its occurrence increases with increasing precipitation and continentality. Common, often dominant, in semi-open forests on water-receiving (floodplain, seepage, and stream-edge) and water-collecting sites; occasional on water-shedding sites when soils are calcareous. Typically associated with ferns (*Athyrium filix-femina*, *Gymnocarpium dryopteris*, and *Polystichum munitum*) and forbs (*Actaea rubra*, *Galium triflorum*, *Tiarella trifoliata*, and *T. unifoliata*). A nitrophytic species characteristic of Moder and Mull humus forms.

Orthilia secunda
Ramischia secunda, Pyrola secunda
One-sided wintergreen

Pyrolaceae
(Wintergreen family)

A shade-tolerant, submontane to subalpine, circumpolar forb (transcontinental in North America). Occurs on moderately dry to fresh, nitrogen-poor soils within boreal, cool temperate, and cool mesothermal climates; its occurrence increases with increasing latitude and precipitation. Widespread in closed-canopy coniferous forests on water-shedding sites. Usually associated with *Hylocomium splendens, Paxistima myrsinites, Rhytidiopsis robusta, Vaccinium alaskaense, V. membranaceum,* and *V. ovalifolium*). An oxylophytic species characteristic of Mor humus forms.

Osmorhiza chilensis
Mountain sweet-cicely

Apiaceae
(Parsley family)

A shade-tolerant/intolerant, submontane to subalpine, South American and transcontinental North American forb. Occurs on fresh to very moist, nitrogen-rich soils within boreal, temperate, and cool mesothermal climates. Widespread but scattered in forest understories on water-receiving (alluvial, floodplain, seepage, and stream-edge) sites, often with a fluctuating groundwater table. Usually associated with *Achlys triphylla, Adenocaulon bicolor, Polystichum munitum, Ribes lacustre, Streptopus roseus,* and *Tiarella trifoliata.* A nitrophytic species characteristic of Moder and Mull humus forms.

Parnassia fimbriata
Fringed grass-of-Parnassus

Parnassiaceae
(Grass-of-parnassus family)

A shade-intolerant, montane to alpine, Western North American forb distributed equally in the Pacific and Cordilleran regions. Occurs in boreal and cool temperate climates on very moist to wet, calcium-rich and nitrogen-rich soils. Common in open-canopy, high-elevation forests, subalpine meadows on water-receiving sites; occasional on water-collecting sites (fens and swamps). Characteristic of Moder and Mull humus forms.

Paxistima myrsinites
Pachistima myrsinites
Falsebox

Celastraceae
(Stafftree family)

A shade-tolerant/intolerant, montane to subalpine, Western North American evergreen shrub distributed more in the Cordilleran than the Pacific region; its occurrence increases with increasing continentality. Occurs predominantly in continental boreal and cool temperate climates on moderately dry to fresh, nitrogen-poor soils. Edaphically comparable to *Gaultheria shallon*. Scattered to plentiful in coniferous forests on water-shedding sites in the coast-interior ecotone, sporadic in maritime climates. Persists on cutover sites; on nutrient-rich sites inhabits decaying wood. Usually associated with *Hylocomium splendens* and *Pleurozium schreberi*. An oxylophytic species characteristic of Mor humus forms.

Pedicularis racemosa
Sickletop lousewort
Pedicularis bracteosa
Bracted lousewort

Scrophulariaceae
(Figwort family)

Shade-intolerant (*P. bracteosa*) to shade-tolerant/intolerant (*P. racemosa*), subalpine, Western North American forbs distributed more in the Cordilleran than the Pacific region. These species occur in continental boreal and cool temperate climates on moderately dry to fresh (*P. racemosa*) or very moist to wet (*P. bracteosa*) nitrogen-medium soils; their occurrence increases with continentality. Common in subalpine meadows and open-canopy, high-elevation forests on water-shedding (*P. racemosa*) and water-receiving sites (*P. bracteosa*) in the coast-interior ecotone.

Pellia neesiana

Pelliaceae

A shade-tolerant, submontane to subalpine, European and North American liverwort distributed equally in the Pacific, Cordilleran, and Atlantic regions. Occurs on very moist to wet, nitrogen-rich (Moder and Mull humus form) within boreal, wet cool temperate, and cool mesothermal climates. Plentiful on water-receiving and water-collecting sites. Usually associated with *Lysichitum americanum*. Characteristic of nutrient-rich wetlands.

Peltigera aphthosa
Peltigera canina
Peltigera membranacea

Peltigeraceae

Shade-tolerant/intolerant, submontane to subalpine, circumpolar lichens (transcontinental in North America). Occur on very dry to moderately dry, nitrogen-poor soils (Mor humus forms) within boreal, cool temperate, and cool mesothermal climates . Widespread in coniferous forests on watershedding sites, occasional on exposed mineral soils and coarse fragments or bedrock. Usually associated with *Pleurozium schreberi* and *Rhacomitrium canescens*. *P. canina* and *P. membranacea* contain nitrogen-fixing blue-green algae. Oxylophytic species characteristic of moisture-deficient sites.

Perideridia gairdneri
Yampah

Apiaceae
(Parsley family)

A very shade-intolerant, submontane to montane, North American forb distributed equally in the Pacific, Cordilleran, and Atlantic regions. Occurs in maritime to submaritime summer-dry cool mesothermal climates on moderately dry to fresh, nitrogen-medium soils; its occurrence increases with increasing temperature and decreases with increasing precipitation. Scattered in forest openings and non-forested communities on water-shedding sites. Characteristic of moisture-deficient sites.

Petasites palmatus

Petasites frigidus var. *palmatus*
Palmate coltsfoot

Petasites frigidus
Sweet coltsfoot

Asteraceae
(Aster family)

P. palmatus - a shade-tolerant/intolerant to very shade intolerant, submontane to subalpine, Asian and transcontinental North American forb. *P. frigidus* - a very shade-intolerant, montane to alpine, circumpolar forbs distributed in Pacific, Cordilleran, and Central North America.

Both species occur on very moist to wet, nitrogen-rich soils within alpine tundra and subalpine boreal climates (*P. frigidus*) or boreal, cool temperate and cool mesothermal climates (*P. palmatus*). Frequent in non-forested communities, and in open-canopy forests, common on floodplains and on exposed mineral soil at roadsides and landslides. Occasional in nutrient-rich wetlands

---

Phegopteris connectilis

Thelypteris phegopteris
Beech fern

Thelypteridaceae

A shade-tolerant/intolerant, montane to subalpine, circumpolar fern (transcontinental in North America). Occurs fresh to very moist, calcium and nitrogen-rich soils within boreal, wet temperate, and cool mesothermal climates; its occurrence increases with precipitation. Rare to scattered in herbaceous understories on water-shedding and water-receiving (floodplain) sites. Fairly common on Queen Charlotte Islands, locally on coastal mainland. A nitrophytic species characteristic of Moder and Mull humus forms.

Philadelphus lewisii
Mock-orange

Hydrangeaceae
(Hydrangea family)

A shade-intolerant, submontane to subalpine, North American deciduous shrub distributed more in the Cordilleran than the Pacific region. Occurs in continental cool temperate and cool semiarid climates on moderately dry to fresh, nitrogen-medium soils; its occurrence increases with increasing temperature and continentality and decreases with increasing elevation. Scattered in open-canopy Douglas-fir forests; in the coast-interior ecotone more frequent on water-shedding sites than water-receiving sites. Characteristic of dry continental forests.

Philonotis fontana

Bartramiaceae

A shade-intolerant, submontane to subalpine, circumpolar moss (transcontinental in North America). Occurs on wet to very wet, nitrogen-medium soils within tundra, boreal, cool temperate, and cool mesothermal climates. Scattered in non-forested, semi-terrestrial communities on water-receiving and water-collecting sites, less frequent in depressions in open-canopy stands where water collects or spring runoff occurs. Characteristic of wetlands.

Phyllodoce empetriformis
Pink mountain-heather
Phyllodoce glanduliflora
P. aleutica ssp. *glanduliflora*
Yellow mountain-heather

Ericaceae
(Heath family)

P. empetriformis - a shade-intolerant, sub-alpine to alpine, oxylophytic, Asian and Western North American evergreen shrub; distributed equally in the Pacific and Cordilleran regions. *P. glanduliflora* - a very shade-intolerant, subalpine to alpine, Western North American evergreen shrub; distributed equally in the Pacific and Cordilleran regions.

Both species occur on moderately dry to fresh, nitrogen-poor soils in alpine tundra and subalpine boreal climates. Common, frequently dominant, in high-elevation, parkland forests and heath communities on water-shedding sites which are often affected by seepage from late-melting snowbanks. Characteristic of alpine communities.

Physocarpus capitatus
Pacific ninebark

Rosaceae
(Rose family)

A shade-tolerant/intolerant, Western North American deciduous shrub distributed more in the Pacific than the Cordilleran region. Occurs in wet cool temperate and cool mesothermal climates on very moist to wet, nitrogen-rich soils (Moder and Mull humus forms); its occurrence decreases with increasing elevation and continentality. Scattered in semi-open or open-canopy forests on water-receiving and water-collecting sites, typically on fine-textured, gleyed alluvial soils with fluctuating ground-water table. Usually associated with *Cornus sericea* and *Rubus spectabilis*. Characteristic of alluvial floodplain forests.

Picea engelmannii
Engelmann spruce

Pinaceae
(Pine family)

A shade-tolerant/intolerant to very shade-intolerant, montane to subalpine, Cordilleran North American evergreen conifer. Grows on a wide range of sites in continental subalpine and montane boreal climates; its occurrence increases with increasing elevation and continentality. Most productive on fresh and moist, nutrient-very rich soils within cool temperate climates. Scattered to abundant in the coast-interior ecotone, often hybridizes with Sitka spruce. Typically associated with lodgepole pine, Pacific silver fir or subalpine fir and white spruce. Because of good survival and productive growth, Engelmann spruce has been preferred over other tree species in regenerating high-elevation sites on eastern Vancouver Island and in the coast-interior ecotone. Characteristic of continental subalpine forests.

Picea sitchensis
Sitka spruce

Pinaceace
(Pine family)

A shade-intolerant, submontane to montane, Pacific North American evergreen conifer. Occurs in hypermaritime to maritime cool mesothermal climates on nitrogen-rich soils. Avoids moisture-deficient and nutrient-deficient soils. Its occurrence increases with increasing latitude and precipitation and decreases with increasing elevation and continentality. Forms pure, open-canopy stands along the outer coast on sites affected by ocean spray and brackish water, and in advanced stages of primary succession on floodplains. Usually associated with black cottonwood, western hemlock, or western redcedar. Most productive on fresh and moist, nutrient-very rich soils within very wet cool mesothermal climates. Characteristic of hypermaritime mesothermal forests.

Pilophoron clavatus   *Stereocaulaceae*

A shade-tolerant/intolerant, submontane to subalpine, Asian and Pacific North American lichen. Occurs in maritime to submaritime subalpine boreal and summer-wet cool meso-thermal climates on very shallow soils. Sporadic on base-poor, exposed coarse fragments and rocks on water-shedding sites. An oxylophytic species characteristic of maritime forests.

Pinus albicaulis   *Pinaceae*
Whitebark pine   (Pine family)

A shade-intolerant, subalpine, Western North American evergreen conifer distributed more in the Pacific than the Cordilleran region. Occurs in continental alpine tundra and subalpine boreal climates on moderately dry to fresh, nitrogen-medium soils; its occurrence increases with increasing continentality and decreases with increasing latitude. Common in parkland forests on water-shedding sites which are free of snow early in the year. Often grows with Engelmann spruce and subalpine fir in the coast-interior ecotone. Characteristic of continental subalpine forests.

Pinus ponderosa
Ponderosa pine

Pinaceae
(Pine family)

A very shade-intolerant, montane, North American evergreen conifer distributed in the Pacific, Cordilleran, and Central regions (rare in the coastal region of British Columbia). Occurs in cool temperate and cool semiarid climates on very dry to moderately dry, nitrogen-medium soils; its occurrence decreases with increasing latitude, elevation, and precipitation. Grows in pure or mixed-species stands (usually with Douglas-fir) on calcium-rich, water-shedding sites in the southern coast-interior ecotone. Most productive on fresh and nutrient-rich soils within cool temperate climates. Characteristic of dry continental forests.

Plagiochila porelloides

Plagiochilaceae

A shade-tolerant, submontane to subalpine, South American and circumpolar liverwort (transcontinental in North America). Occurs on fresh to very moist, nitrogen-medium soils within boreal, wet cool temperate, and cool mesothermal climates; its occurrence increases with increasing precipitation. Scattered on organic substrates in closed-canopy forests on water-receiving sites; occasional on coarse fragments, rock and bark. Characteristic of calcium-rich substrates.

Plagiomnium insigne   *Mniaceae*
Mnium insigne

A shade-tolerant, submontane to montane, Western North American moss distributed more in the Pacific than the Cordilleran region. Occurs in cool mesothermal climates on very moist to wet, nitrogen-rich soils; its occurrence decreases with increasing elevation and continentality. Common on water-receiving sites; inhabits exposed mineral soils, occasionally calcium-rich bark. Usually associated with *Leucolepis menziesii*, *Polystichum munitum*, and *Tiarella trifoliata*. Characteristic of Moder and Mull humus forms.

Plagiothecium undulatum   *Plagiotheciaceae*

A shade-tolerant, submontane to montane, circumpolar moss distributed mainly in Pacific North America; its occurrence decreases with increasing precipitation, continentality, and elevation. Occurs in hypermaritime to maritime cool mesothermal climates on fresh to very moist, nitrogen-poor soils. Scattered to plentiful in closed-canopy, old-growth, coniferous forests on water-shedding and water-receiving sites. Common on decaying coniferous wood, rare on base-poor coarse fragments and rocks. Usually associated with *Rhytidiadelphus loreus*, *Vaccinium alaskaense*, and *V. ovalifolium*. An oxylophytic species characteristic of Mor humus forms.

Platanthera dilatata
Habenaria dilatata
White bog-orchid

Orchidaceae
(Orchid family)

A shade-intolerant, submontane to subalpine, Asian and transcontinental North American forb. Occurs on wet to very wet, nitrogen-medium soils within boreal, temperate, and cool mesothermal climates; its occurrence increases with increasing precipitation and continentality. Scattered on organic substrates in non-forested, semi-terrestrial communities and open-canopy forests on water-collecting sites. Characteristic of wetlands.

Platanthera orbiculata
Habenaria orbiculata
Round-leaved rein-orchid

Orchidaceae
(Orchid family)

A shade-tolerant, montane to subalpine, transcontinental North American forb (rare in the Pacific region). Occurs in continental boreal and cool temperate climates on moderately dry to fresh, nitrogen-poor soils; its occurrence increases with increasing continentality. Scattered in the mossy understory of semi-open coniferous forests on water-shedding sites in the coast-interior ecotone. Usually associated with *Clintonia uniflora*, *Pleurozium schreberi*, *Ptilium crista-castrensis*, *Rhytidiopsis robusta*, and *Vaccinium membranaceum*. An oxylophytic species characteristic of Mor humus forms.

Pleurozium schreberi   *Entodontaceae*

A shade-tolerant/intolerant, submontane to alpine, South American and circumpolar moss (transcontinental in North America); sporadic in the Pacific region). Occurs predominantly in continental boreal and cool temperate climates on nitrogen-poor soils; its occurrence increases with increasing continentality. Widespread, frequently dominant, in semi-open coniferous forests on water-shedding sites in the coast-interior ecotone. Occasional on drying surfaces on water-receiving and on water-collecting sites, and on exposed mineral soil and coarse fragments or rocks. In maritime climates associated with *Hylocomium splendens*, *Gaultheria shallon*, and *Vaccinium membranaceum*. In continental climates associated with *Paxistima myrsinites*, *Ptilium crista-castrensis*, and *Vaccinium membranaceum*. An oxylophytic species characteristic of Mor humus forms.

Pogonatum alpinum   *Polytrichaceae*
Polytrichum alpinum
Polytrichum macounii

A shade-tolerant, submontane to subalpine, Asian and North American moss distributed more in the Pacific than the Cordilleran region. Occurs in cool temperate and cool mesothermal climates on fresh to very moist, nitrogen-medium soils. Its occurrence decreases with increasing elevation, temperature, and continentality. Sporadic in closed-canopy coniferous forests on water-receiving sites. Characteristic of friable Mor and acidic Moder humus forms.

Pogonatum contortum   *Polytrichaceae*

A shade-tolerant, submontane to subalpine, Asian and Pacific North American moss. Occurs in cool mesothermal climates on fresh to very moist, nitrogen-medium soils; its occurrence decreases with increasing elevation, continentality, and temperature. Sporadic on exposed mineral soils in closed-canopy forests and exposed mineral soils. Characteristic of microsites with friable mineral soil churned by windthrow.

Polypodium glycyrrhiza   *Polypodiaceae*
Licorice fern   (Polypody family)

A shade-intolerant, submontane to montane, Asian and Pacific North American fern. Occurs in maritime to submaritime cool mesothermal climates on very shallow, calcium-rich soils; its occurrence decreases with increasing continentality and elevation. Scattered to plentiful in open-canopy forests on water-shedding and water-receiving sites. Inhabits exposed soils, coarse fragments (boulders), or cliffs affected by temporary surface flow of water and/or fine organic materials. Common on the calcium-rich bark of broad-leaved trees, typically on *Acer macrophyllum*. Characteristic of mesothermal forests.

Polypodium scouleri
Leatherleaf polypody

Polypodiaceae
(Polypody family)

A shade-intolerant, submontane to montane, Pacific North American fern. Occurs in hypermaritime to maritime wet cool mesothermal climates on very shallow soils; its occurrence decreases with increasing elevation and continentality. Common in ocean spray-affected (littoral), open-canopy forests in the proximity of the Pacific Ocean. As does *P. glycyrrhiza*, *P. scouleri* inhabits exposed mineral soils, coarse fragments (boulders), and cliffs affected by temporary surface flow of water and fine organic materials. Common on stems and branches of trees, typically on *Picea sitchensis*. Characteristic of hypermaritime mesothermal forests.

Polystichum braunii
Braun's holly-fern

Aspleniaceae
(Spleenwort family)

A shade-tolerant/intolerant, submontane to subalpine, circumpolar fern (transcontinental in North America but rare in the Central region). Occurs on fresh to very moist, nitrogen-rich soils within subalpine boreal, cool temperate, and cool mesothermal climates. Occurs sparsely in semi-open stands on water-receiving sites; most common on colluvial soils enriched by temporary seepage and surface flow of fine organic materials. A nitrophytic species characteristic of Moder and Mull humus forms.

Polystichum lonchitis
Mountain holly-fern

Aspleniaceae
(Spleenwort family)

A shade-intolerant, montane to alpine, circumpolar fern (transcontinental in North America). Occurs in alpine tundra and subalpine boreal climates on moderately dry to fresh, nitrogen-medium soils; its occurrence increases with increasing elevation. Sporadic in non-forested and open-canopy forested communities on water-shedding sites; most common calcium-rich colluvial (scree and talus) soils. Inhabits friable organic materials accumulated between base-rich coarse fragments (stones and boulders) and on cliffs affected by temporary surface flow of water and organic materials. Characteristic of colluvial sites.

Polystichum munitum
Sword fern

Aspleniaceae
(Spleenwort family)

A shade-tolerant/intolerant, submontane to subalpine, Western North American fern distributed more in the Pacific than the Cordilleran region. Occurs in cool mesothermal climates on nitrogen-rich soils; its occurrence decreases with increasing elevation and continentality. Widespread in forest understories. Persists on cutover sites, sporadic to scattered on water-shedding sites, plentiful to abundant (frequently dominant) on water-receiving and colluvial sites enriched by surface flow of fine organic materials. Commonly associated with *Achlys triphylla*, *Mahonia nervosa*, and *Tiarella trifoliata*. A nitrophytic species characteristic of Moder and Mull humus forms.

Polytrichum juniperinum   *Polytrichaceae*

A shade-tolerant/intolerant, submontane to alpine, cosmopolitan moss (transcontinental in North America). Grows on water-shedding sites in tundra, boreal, cool temperate, cool semiarid, and cool mesothermal climates; its occurrence decreases with increasing elevation and increases with increasing continentality. Scattered in open-canopy forests on strongly drained soils on rock outcrops where it is commonly associated with *Gaultheria shallon*. Plentiful to abundant in early-seral communities on cutover and burnt sites where it grows with *Epilobium angustifolium*. Also inhabits exposed sandy soils. Characteristic of fire-disturbed sites.

Polytrichum piliferum   *Polytrichaceae*

A shade-tolerant/intolerant, submontane to alpine, cosmopolitan moss (transcontinental in North America). Occurs on excessively dry to very dry, nitrogen-poor soils within tundra, boreal, cool temperate, cool semiarid, and cool mesothermal climates. Its occurrence decreases with increasing elevation and increases with increasing continentality. Scattered in open-canopy coniferous forests on strongly drained water-shedding sites where it inhabits shallow forest floors or exposed sandy soils. Commonly associated with lichens and *Gaultheria shallon*. Characteristic of moisture-deficient sites.

Populus tremuloides
Trembling aspen

Salicaceae
(Willow family)

A shade-intolerant, submontane to alpine, transcontinental North American, deciduous broad-leaved tree (rare in the Pacific region). Occurs in continental boreal and cool temperate climates on nitrogen-rich soils (Moder and Mull humus forms). Its occurrence increases with increasing continentality and latitude; sporadic in the coast-interior ecotone. Forms pure stands on base-rich, water-shedding and water-receiving sites following disturbance. A nurse species, it improves humus form quality through nutrient-rich litterfall. Regenerates abundantly from root suckers which quickly form a dense shrub layer and suppress the growth of shade-intolerant conifers. Characteristic of young-seral, continental forests.

Populus trichocarpa
Populus balsamifera ssp. *trichocarpa*
Black cottonwood

Salicaceae
(Willow family)

A shade-intolerant, submontane to montane, Western North American deciduous broad-leaved tree. Occurs on water-receiving sites within boreal, temperate, cool semiarid, and mesothermal climates; its occurrence decreases with increasing elevation. Grows on fresh to very moist, nitrogen-rich soils (Moder and Mull humus forms); tolerates a fluctuating groundwater table. Forms pure stands in primary succession on floodplains and stream-edge sites. Its calcium-rich bark supports some bryophyte communities. This fast-growing tree regenerates abundantly from root sprouts, stump sprouts, and cuttings (buried branches) on disturbed sites, thus suppressing growth of shade-intolerant conifers. A nitrophytic species characteristic of alluvial floodplain forests.

Potentilla glandulosa
Sticky cinquefoil

Rosaceae
(Rose family)

A shade-intolerant, submontane to subalpine, Western North American forb distributed equally in the Pacific and Cordilleran regions. Occurs on moderately dry to fresh, nitrogen-medium soils within temperate, cool semiarid, and mesothermal climates. Its occurrence increases with increasing temperature and continentality and decreases with increasing latitude. Characteristic of non-forested communities and open-canopy forests on water-shedding sites.

Prenanthes alata
Western rattlesnake-root

Asteraceae
(Aster family)

A shade-tolerant/intolerant, submontane to montane, Pacific North American forb. Occurs in maritime to submaritime wet cool mesothermal climates on fresh to moist, nitrogen-rich soils (Moder and Mull humus forms); its occurrence decreases with increasing elevation and continentality. Grows on water-receiving and water-collecting sites (usually stream-edge sites) where it inhabits recently deposited mineral materials. Sporadic in open-canopy broad-leaved forests; often associated with *Adiantum pedatum, Aruncus dioicus*, and *Oplopanax horridus* on flooded sites. A nitrophytic species characteristic of Moder and Mull humus forms.

Prunus virginiana
Padus virginiana
Choke cherry

Rosaceae
(Rose family)

A shade-intolerant, submontane to subalpine, transcontinental North American deciduous shrub. Occurs in continental cool temperate and cool semiarid climates on moderately dry to fresh, nitrogen-rich (occasionally weakly alkaline) soils (Moder and Mull humus forms). Its occurrence increases with increasing temperature and continentality, and decreases with increasing precipitation and latitude. Occasional in immature broadleaf forests on water-shedding sites on leeward Vancouver Island and in the coast-interior ecotone. A nitrophytic species characteristic of young-seral, continental forests.

Pteridium aquilinum
Bracken

Dennstaediaceae
(Hay-scented fern family)

A shade-tolerant/intolerant, submontane to subalpine, cosmopolitan fern (transcontinental in North America). Occurs on water-shedding and water-receiving sites in boreal, temperate, mesothermal, and tropical climates; its occurrence decreases with increasing latitude and elevation. Scattered in coniferous forests where it is usually associated with *Gaultheria shallon*, *Hylocomium splendens*, and *Vaccinium parvifolium*; plentiful to abundant, often dominant, in initial communities on cutover and burnt sites where its vigorous growth and litterfall may hinder growth of conifers. Characteristic of fire-disturbed sites.

Pterospora andromeda
Pinedrops

Monostropaceae
(Indian-pipe family)

A shade-tolerant/intolerant, submontane to subalpine, circumpolar saprophyte (transcontinental in North America). Occurs on very dry to moderately dry, nitrogen-medium soils within boreal, temperate, cool semiarid, and mesothermal climates. Its occurrence increases with increasing continentality and decreases with increasing elevation. Infrequent in the mossy understory of coniferous (Douglas-fir and lodgepole pine) forests on water-shedding sites. Usually associated with *Arctostaphylos uva-ursi, Hylocomium splendens, Pleurozium schreberi, Rosa acicularis,* and *Spiraea betulifolia.* Characteristic of moisture-deficient sites.

Ptilium crista-castrensis

Hypnaceae

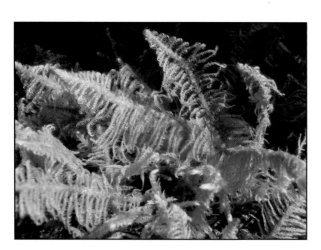

A shade-tolerant, montane to subalpine, circumpolar moss (transcontinental in North America). Occurs in continental boreal and cool temperate climates on nitrogen-poor soils; its occurrence increases with increasing latitude, continentality, and precipitation. Common and frequently dominant in closed-canopy coniferous forests on water-shedding and water-receiving sites; often inhabits decaying wood. Usually associated with *Clintonia uniflora, Cornus canadensis, Hylocomium splendens, Paxistima myrsinites, Pleurozium schreberi,* and *Vaccinium membranaceum* in the coast-interior ecotone. An oxylophytic species characteristic of Mor humus forms.

Pyrola asarifolia, Pink wintergreen
P. chlorantha, Green wintergreen
P. virens
P. picta
White-veined wintergreen

Pyrolaceae
(Wintergreen family)

P. asarifolia and *P. chlorenta* - shade-tolerant/ intolerant, submontane to subalpine, circumpolar forbs (transcontinental in North America). Grows in boreal, temperate, and cool mesothermal climates; its occurrence increases with increasing latitude and decreases with increasing elevation. *P. picta* - a shade-tolerant/intolerant, submontane to montane, Western North American forb; its occurrence decreases with increasing elevation and latitude.

These species occur on moderately dry to fresh, nitrogen-medium soils. Widespread and scattered in coniferous forests on water-shedding sites (*P. asarifolia* often in alluvial forests). Usually associated with *Hylocomium splendens*, *Linnaea borealis*, and *Pleurozium schreberi*. Characteristic of friable Mor and acidic Moder humus forms.

Quercus garryana
Garry oak

Fagaceae
(Beech family)

A shade-intolerant, submontane to montane, Western North American deciduous broadleaved tree distributed more in the Pacific than the Cordilleran region. Occurs in maritime to submaritime summer-dry cool mesothermal climates on very dry to moderately dry soils; its occurrence decreases with increasing latitude, elevation, precipitation, and continentality. Sporadic to scattered on southeastern Vancouver Island and islands of Georgia Strait (less frequent on adjacent mainland). Usually forms open-canopy stands on water-shedding (rock outcrops) sites. Its calcium-rich bark supports corticolous moss communities. Characteristic of dry mesothermal forests.

Ranunculus eschscholtzii
Subalpine buttercup

Ranunculaceae
(Buttercup family)

A very shade-intolerant, sub-alpine to alpine, Asian and Western North American forb distributed equally in the Pacific and Cordilleran regions. Occurs in alpine tundra and subalpine boreal climates on very moist to wet, nitrogen-rich soils (Moder and Mull humus forms); its occurrence increases with increasing latitude. Common in non-forested, meadow-like and stream-edge communities on seepage or flooded sites. Characteristic of alpine and subalpine communities.

Ranunculus occidentalis
Western buttercup

Ranunculaceae
(Buttercup family)

A shade-intolerant, submontane to subalpine, Asian and Western North American forb distributed more in the Pacific than the Cordilleran region. Occurs on nitrogen-medium (often disturbed) soils within subalpine boreal, montane boreal, and cool mesothermal climates. Scattered to plentiful and occasionally dominant in early-seral, herbaceous (subalpine meadows) and grassy communities on water-shedding or water-receiving sites. Characteristic of disturbed sites.

Ranunculus repens
Creeping buttercup

Ranunculaceae
(Buttercup family)

A shade-intolerant, submontane to montane, Eurasian forb introduced to North America where it is transcontinental. Occurs on very moist to wet, nitrogen-rich soils; its occurrence decreases with increasing elevation within montane boreal, wet temperate, and cool mesothermal climates. A widespread weed in non-forested, early-seral communities on disturbed water-receiving and water-collecting sites; most frequent on exposed mineral soil along streams. Tolerates flooding and fluctuating groundwater tables. A nitrophytic species characteristic of Moder and Mull humus forms.

Ranunculus uncinatus
Ranunculus bongardii
Little buttercup

Ranunculaceae
(Buttercup family)

A shade-intolerant, submontane to subalpine, Asian and North American forb distributed more in the Pacific than the Cordilleran region. Occurs on fresh to very moist, nitrogen-rich soils within boreal, wet temperate, and cool mesothermal climates; its occurrence decreases with increasing elevation. Tolerates flooding and fluctuating groundwater tables. Occasional on exposed mineral soils in forest openings or open-canopy broad-leaved forests on water-receiving sites; commonly inhabits recently deposited organic matter-rich alluvium. A nitrophytic species characteristic of Moder and Mull humus forms.

Rhacomitrium canescens
Rhacomitrium heterostichum

Grimmiaceae

Shade-intolerant, submontane to alpine, circumpolar mosses (transcontinental in North America). Occur on excessively dry to very dry, nitrogen-poor soils (Mor humus forms) within tundra, boreal, cool temperate, and cool mesothermal climates; their occurrence decreases with increasing elevation and precipitation. Common in non-forested communities and open-canopy coniferous forests on rock outcrops and coarse-skeletal to fragmental rapidly drained alluvial, colluvial, and glaciofluvial materials. Occasional in nutrient-poor wetlands on topographic prominences subjected to regular desiccation. Typically associated with lichens. Oxylophytic species characteristic of moisture-deficient sites.

Rhamnus purshianus
Cascara

Rhamnaceae
(Buckthorn family)

A shade-tolerant/intolerant, submontane to subalpine, Western North American deciduous shrub distributed more in the Pacific than the Cordilleran region. Occurs in wet cool temperate and cool mesothermal climates on very moist to wet, nitrogen-rich soils; its occurrence decreases with increasing elevation, latitude, and continentality. Tolerates fluctuating groundwater tables. Scattered in semi-open and open-canopy disturbed forests on water-receiving and water-collecting sites; occasional on water-shedding sites. Often associated with *Acer circinatum, Athyrium filix-femina,* and *Polystichum munitum.* Characteristic of Moder and Mull humus forms.

Rhizomnium glabrescens
Mnium glabrescens

Mniaceae

A very shade-tolerant, submontane to sub-alpine, Western North American moss distributed more in the Pacific than the Cordilleran region. Occurs in cool mesothermal climates on fresh to very moist, nitrogen-medium soils; its occurrence increases with increasing precipitation and decreases with increasing continentality and elevation. Scattered to abundant in old, closed-canopy coniferous forests on water-shedding and water-receiving sites; occasional on the decaying wood and bark of coniferous trees. Usually associated with *Blechnum spicant*, *Dryopteris expansa*, and *Rhytidiadelphus loreus*. Characteristic of friable Mor and acidic Moder humus forms.

Rhizomnium magnifolium
Mnium magnifolium

Mniaceae

A shade-tolerant, submontane to alpine, circumpolar moss (transcontinental in North America). Occurs on wet to very wet, nitrogen-rich soils (Moder and Mull humus forms) within boreal, cool temperate, and cool mesothermal climates; its occurrence increases with increasing precipitation and decreases with increasing continentality. Scattered to plentiful in closed-canopy, coniferous forests on water-collecting sites; usually associated with *Lysichitum americanum*. Characteristic of nutrient-rich wetlands.

Rhizomnium nudum
Mnium nudum

Mniaceae

A shade-tolerant, submontane to alpine, Western North American moss distributed more in the Pacific than the Cordilleran region. Occurs in maritime subalpine boreal and cold mesothermal climates on very moist to wet, nutrient-medium soils. Its occurrence increases with increasing elevation and precipitation and decreases with increasing continentality. Common in closed-canopy, high-elevation coniferous forests on water-receiving and water-collecting sites. Usually associated with *Blechnum spicant, Streptopus roseus*, and *Veratrum eschsscholtzii*. Characteristic of subalpine forests on late snow-melt sites.

Rhododendron albiflorum
White-flowered rhododendron

Ericaceae
(Heath family)

A shade-tolerant/intolerant, montane to alpine, Western North American deciduous shrub distributed more in the Cordilleran than the Pacific region. Occurs predominantly in continental subalpine boreal climates on moderately dry to fresh, nitrogen-poor soils (Mor humus forms); its occurrence increases with increasing continentality and decreases with increasing latitude. Common and often dominant on water-shedding sites in open-canopy coniferous forests in the coast-interior ecotone; often on decaying coniferous wood. Typically associated with *Barbilophozia floerkei, B. lycopodioides, Rhytidiopsis robusta*, and *Vaccinium membranaceum*. An oxylophytic species characteristic of subalpine continental forests.

Rhynchospora alba
Beak-rush, beak-sedge

Cyperaceae
(Sedge family)

A very shade-intolerant, sub-montane to subalpine, circum-polar sedge [transcontinental in North America (absent in Alberta and Manitoba)]. Occurs on wet to very wet, nitrogen-poor soils (Mor humus forms) within boreal, cool temperate, and cool mesothermal climates; its occurrence increases with increasing precipitation. Scattered in non-forested, semi-terrestrial communities on water-collecting sites; occasional on acidic sands. An oxylophytic species characteristic of nutrient-poor wetlands.

Rhytidiadelphus loreus

Rhytidiaceae

A very shade-tolerant, submontane to sub-alpine, European and North American moss distributed in the Pacific, Cordilleran (sporadically), and Atlantic regions. One of the most common moss species in coastal British Columbia. Occurs in hypermaritime to submaritime subalpine boreal and cool mesothermal climates on fresh to moist, nitrogen-poor soils; its occurrence increases with increasing precipitation, and decreases with increasing elevation and continentality. Plentiful to abundant in closed-canopy coniferous forests on water-shedding and water-receiving sites; on nutrient-rich sites inhabits decaying coniferous wood. Usually associated with *Blechnum spicant*, *Dryopteris expansa*, *Plagiothecium undulatum*, *Vaccinium alaskaense*, and *V. parvifolium*. An oxylophytic species characteristic of Mor humus forms.

Rhytidiadelphus triquetrus   *Rhytidiaceae*

A shade-tolerant/intolerant, submontane to subalpine, circumpolar moss (transcontinental in North America). Occurs on nitrogen-medium soils, within boreal, cool temperate, and cool mesothermal climates; its occurrence increases with increasing temperature and continentality and decreases with increasing elevation. Common and occasionally dominant in mossy or herbaceous forest understories on water-shedding and water-receiving sites; occasionally inhabits decaying wood. Usually associated with *Acer glabrum*, *Disporum hookeri*, *Mahonia aquifolium*, *M. nervosa*, and *Kindbergia oregana*. Characteristic of friable Mor and acidic Moder humus forms.

Rhytidiopsis robusta   *Rhytidiaceae*

A very shade-tolerant, montane to alpine, Western North American moss distributed equally in the Pacific and Cordilleran regions. Occurs in boreal and cool temperate climates on nitrogen-poor soils; its occurrence increases with increasing elevation and continentality. Plentiful to abundant (often dominant) in subalpine forests on water-shedding sites; occasionally inhabits decaying coniferous wood. Typically associated with *Menziesia ferruginea*, *Rhododendron albiflorum*, and *Vaccinium ovalifolium* and *V. membranaceum*. An oxylophytic species characteristic of Mor humus forms.

Ribes bracteosum
Stink currant

Grossulariaceae
(Currant or Gooseberry family)

A shade-tolerant/intolerant, submontane to subalpine, Western North American deciduous shrub distributed more in the Pacific than in the Cordilleran region. Occurs in hypermaritime to maritime cool mesothermal climates on very moist to wet, nitrogen-rich soils; its occurrence increases with increasing precipitation and decreases with increasing elevation and continentality. Scattered in semi-open forests on water-receiving (floodplain and stream-edge) sites. Usually associated with *Alnus rubra*, *Oplopanax horridus*, and *Rubus spectabilis*. A nitrophytic species characteristic of Moder and Mull humus forms.

Ribes divaricatum
Wild gooseberry

Grossulariaceae
(Currant or Gooseberry family)

A shade-tolerant/intolerant, submontane to montane, Western North American deciduous shrub distributed more in the Pacific than in the Cordilleran region. Occurs in maritime to submaritime cool mesothermal climates on moderately dry to fresh, nitrogen-medium soils; its occurrence decreases with increasing elevation and continentality. Sporadic in semi-open Douglas-fir forests on water-shedding sites; persists on cutover areas. Characteristic of early-seral communities.

Ribes lacustre
Black gooseberry

Grossulariaceae
(Currant or Gooseberry family)

A shade-tolerant/intolerant, submontane to subalpine, Asian and transcontinental North American deciduous shrub. Occurs on fresh to moist, nitrogen-rich soils within boreal, temperate, and cool mesothermal climates; its occurrence increases with increasing continentality. Common in semi-open forests on water-receiving sites, less often on water-shedding sites. Usually associated with *Acer glabrum*, *Oplopanax horridus*, *Rubus parviflorus*, *Smilacina stellata*, and *Valeriana sitchensis*. A nitrophytic species characteristic of Moder and Mull humus forms.

Ribes laxiflorum
Trailing black currant

Grossulariaceae
(Currant or Gooseberry family)

A shade-tolerant/intolerant, submontane to subalpine, Asian and Western North American deciduous shrub distributed equally in the Pacific and Cordilleran regions. Occurs on very moist to wet, nitrogen-rich soils (Moder or Mull humus forms) within boreal, cool temperate, and cool mesothermal climates; its occurrence increases with increasing continentality. Sporadic in semi-open and open-canopy forests on water-receiving and water-collecting sites. A nitrophytic species characteristic of nutrient-rich wetlands.

Ribes sanguineum
Red-flowering currant
Ribes lobbii
Gummy gooseberry

Grossulariaceae
(Currant or Gooseberry family)

Shade-intolerant, submontane to subalpine, Western North American deciduous shrubs distributed more in the Pacific than the Cordilleran region. These species occur in maritime to submaritime cool mesothermal climates on very dry to moderately dry, nitrogen-medium soils; their occurrence decreases with increasing elevation, precipitation, and continentality. Sporadic to scattered in early-seral communities and open-canopy Douglas-fir forests on water-shedding sites. Characteristic of moisture-deficient sites.

Rosa acicularis
Prickly rose

Rosaceae
(Rose family)

A shade-intolerant, montane to subalpine, circumpolar deciduous shrub [transcontinental in North America (rare in the Pacific region)]. Occurs in continental boreal and cool temperate climates on moderately dry to fresh, nitrogen-medium soils. Common in open-canopy forests on water-shedding and water-receiving sites in the eastern coast-interior ecotone. Characteristic of continental forests.

Rosa gymnocarpa
Baldhip rose

Rosaceae
(Rose family)

A shade-tolerant/intolerant, submontane to subalpine, Western North American deciduous shrub distributed equally in the Pacific and Cordilleran regions. Occurs on very dry to moderately dry, nutrient-medium soils within boreal, temperate, and mesothermal climates; its occurrence decreases with increasing precipitation, latitude, and elevation. Common in open-canopy forests on water-shedding sites; persists on cutover areas. Usually associated with *Gaultheria shallon*, *Kindbergia oregana*, and *Mahonia nervosa*. Characteristic of moisture-deficient sites.

Rosa nutkana
Nootka rose

Rosaceae
(Rose family)

A shade-tolerant/intolerant, submontane to montane, Western North American deciduous shrub distributed more in the Pacific than the Cordilleran region. Occurs on fresh to very moist, nitrogen-rich soils within boreal, cool temperate, and cool mesothermal climates. Most frequent on floodplains; sporadic in non-forested communities and open-canopy forests on water-shedding sites with fluctuating groundwater tables. Occasional on sites affected by ocean spray and brackish water. A nitrophytic species characteristic of Moder and Mull humus forms.

Rubus idaeus
Rubus strigosus
Red raspberry

Rosaceae
(Rose family)

A shade-intolerant, submon-tane to subalpine, circumpo-lar deciduous shrub [transcon-tinental in North America (ab-sent in hypermaritime and maritime climates)]. Occurs in continental boreal and wet temperate climates on fresh to very moist, nitrogen-rich soils. Plentiful in early successional communities on cutover and/ or burnt sites in the coast-interior ecotone; scattered in open-canopy stands on water-shedding and water-receiving sites. Usually associated with *Epilobium angustifolium* and *Rubus parviflorus*. May hinder natural regeneration and growth of shade-intolerant conifers. A nitrophytic species characteristic of disturbed sites.

Rubus leucodermis, **Black raspberry**
Rubus laciniatus, **Evergreen blackberry**

Rosaceae
(Rose family)

R. leucodermis - a shade-intol-erant, submontane to sub-alpine, Western North Ameri-can deciduous shrub distrib-uted equally in the Pacific and Cordilleran regions. *R. lacini-atus* - a shade-intolerant, submontane to montane, Eu-ropean deciduous shrub intro-duced to North America (pres-ently transcontinental).

These species occur on wa-ter-shedding and water-receiv-ing sites in boreal, temperate, cool semiarid, and cool meso-thermal climates; on fresh to very moist (*R. laciniatus*) or moderately dry to fresh (*R. leucodermis*), nitrogen-rich soils. Plentiful in initial commu-nities on cutover and burnt sites; scattered in open-canopy, immature forests. Often associated with *Epilobium angustifolium*, *Pteridium aquilinum*, and *Rubus parviflorus*. May hinder natural regeneration, and establishment of shade-intolerant conifers. Nitrophytic species characteristic of disturbed sites.

Rubus parviflorus
Thimbleberry

Rosaceae
(Rose family)

A shade-tolerant/intolerant, submontane to subalpine, North American deciduous shrub distributed equally in the Pacific, Cordilleran, and Central regions. Occurs on nitrogen-rich soils within boreal, temperate, and mesothermal climates; its occurrence decreases with increasing elevation and latitude and increases with increasing continentality. Very common in open-canopy forests and early-seral communities on cutover and/or burnt sites where it may hinder natural regeneration and growth of shade-intolerant conifers. Usually associated with *Alnus rubra*, *Athyrium filix-femina*, *Epilobium angustifolium*, *Oplopanax horridus*, *Rubus spectabilis*, *Sambucus racemosa*, *Streptopus roseus*, and *Tiarella unifoliata*. A nitrophytic species characteristic of Moder and Mull humus forms.

Rubus pedatus
Five-leaved bramble

Rosaceae
(Rose family)

A shade-tolerant, montane to subalpine, Asian and Western North American forb distributed equally in the Pacific and Cordilleran regions. Occurs in boreal and cool temperate climates on fresh to very moist, nitrogen-poor soils; its occurrence increases with increasing elevation and precipitation and decreases with increasing latitude. Common in semi-open coniferous forests on water-shedding and water-receiving sites. Typically associated with *Blechnum spicant*, *Clintonia uniflora*, *Rhododendron albiflorum*, *Rhytidiopsis robusta*, *Vaccinium alaskaense*, and *V. membranaceum*. An oxylophytic species characteristic of Mor humus forms.

Rubus pubescens
Trailing raspberry

Rosaceae
(Rose family)

A shade-tolerant/intolerant, submontane to subalpine, transcontinental North American forb (in the Pacific region, eastern parts of the coast-interior ecotone). Occurs in continental boreal and temperate climates on fresh to very moist, nitrogen-rich soils. Scattered in herbaceous understories on water-receiving sites throughout the interior region. A nitrophytic species characteristic of Moder and Mull humus forms.

Rubus spectabilis
Salmonberry

Rosaceae
(Rose family)

A shade-tolerant/intolerant, submontane to subalpine, Asian and Western North American deciduous shrub distributed more in the Pacific than the Cordilleran region. Occurs in hypermaritime to maritime cool mesothermal climates on very moist to wet, nitrogen-rich soils; its occurrence increases with increasing precipitation and decreases with increasing elevation and continentality. Very common on water-receiving (floodplain and seepage) and water-collecting sites; tolerates fluctuating groundwater tables. Of-

ten dominant in early-seral communities where it hinders natural regeneration and growth of shade-intolerant conifers. Usually associated with *Alnus rubra, Athyrium filix-femina, Lysichitum americanum, Oplopanax horridus, Rubus parviflorus,* and *Tiarella trifoliata.* A nitrophytic species characteristic of Moder and Mull humus forms.

Rubus ursinus
Trailing blackberry

Rosaceae
(Rose family)

A shade-tolerant/intolerant, submontane to montane, Western North American deciduous shrub distributed more in the Pacific than the Cordilleran region. Occurs in maritime to submaritime cool mesothermal climates on moderately dry to fresh, nitrogen-medium soils; its occurrence decreases with increasing elevation and continentality. Common but scattered in forest understories on disturbed, water-shedding sites, often plentiful in disturbed and early seral communities on cutover-and/or burnt sites. Usually associated with *Anaphalis margaritacea, Epilobium angustifolium, Gaultheria shallon, Kindbergia oregana, Mahonia nervosa*, and *Pteridium aquilinum*. Characteristic of young-seral mesothermal forests.

Salix bebbiana
Bebb's willow
Salix sitchensis
Sitka willow

Salicaceae
(Willow family)

A shade-intolerant, montane, Asian and transcontinental North American deciduous shrubs, in the Pacific region *S. bebiana* present only in the coast-interior transition, *S. sitchensis* more in the Pacific rather than the Cordilleran region. Occur in subcontinental to continental boreal and cool temperate climates.

Scattered to plentiful on nitrogen-medium soils, especially in disturbed and early seral communities, common in flood plains. May hinder natural regeneration and growth of shade-tolerant conifers.

Salix scouleriana
Scouler's willow
Salix hookeriana
Hooker's willow

Salicaceae
(Willow family)

Shade-intolerant deciduous shrubs, *S. scouleraina* distributed equally in the Pacific and Cordilleran regions (and marginally in the Central region) within boreal, temperate and cool mesothermal climates. *S. hookeriana* is restricted to the Pacific region.

Occur on nitrogen-medium soils in early stages of primary and secondary succession. *Salix scouleriana* grows in a wide range of soil moisture regimes and may hinder natural regeneration and growth of shade-intolerant conifers. *S. hookeriana* is restricted to very moist to wet soils.

Sambucus racemosa
Sambucus pubens
Red elderberry

Caprifoliaceae
(Honeysuckle family)

A shade-tolerant to shade-tolerant/intolerant, submontane to subalpine, circumpolar deciduous shrub (transcontinental in North America). Occurs on fresh to very moist, nitrogen-rich soils within boreal, temperate, and cool mesothermal climates; its occurrence decreases with increasing elevation. Scattered to plentiful in open-canopy forests on water-receiving sites. Indicative of rapid decomposition of forest floor materials (originally Mor humus forms) remaining on cutover or fire-disturbed, water-shedding sites. Usually associated with *Alnus rubra*, *Athyrium filix-femina*, *Epilobium angustifolium*, *Rubus parviflorus*, and *R. spectabilis*. A nitrophytic species characteristic of Moder and Mull humus forms.

Sanguisorba canadensis
Sanguisorba sitchensis
Sitka burnet
Sanguisorba officinalis
Great burnet

Rosaceae
(Rose family)

S. canadensis - a shade-intolerant, submontane to subalpine, Asian and trans-continental North American forb. *S. officinalis* - a shade-intolerant, submontane to subalpine, circumpolar forb distributed in Pacific and Cordilleran North America.

These species occur on very moist to wet soils, *S. officinalis* on nitrogen-medium, often disturbed soils. Widespread but scattered on water-receiving and water-collecting sites in boreal, cool temperate, and cool mesothermal climates. Most common in non-forested semi-terrestrial, often *Sphagnum*-dominated communities. Characteristic of wetlands.

Sanicula graveolens
Sierra sanicle
Sanicula crassicaulis
Pacific sanicle

Apiaceae
(Parsley family)

S. crassicaulis - a shade-tolerant to shade-intolerant, submontane to montane, South American and Pacific North American forb. *S. graveolens* - a shade-tolerant/intolerant to very shade-intolerant, submontane to subalpine, South American and Western North American forb distributed equally in the Pacific and Cordilleran regions.

Both species occur on very dry to moderately dry, nitrogen-rich soils (Moder or Mull humus forms). Sporadic to scattered in open-canopy, young-seral forests on base-rich, water-shedding sites. Nitrophytic species characteristic of moisture-deficient sites.

Satureja douglasii
Yerba buena

Lamiaceae
(Mint family)

A shade-tolerant/intolerant, submontane to montane, Western North American evergreen (ligneous) forb distributed more in the Pacific than the Cordilleran region. Occurs in maritime to submaritime summer-dry cool meso-thermal climates on moderately dry to fresh, nitrogen-rich soils; its occurrence decreases with increasing elevation, precipitation, and distance from the ocean. Sporadic to scattered in Douglas-fir forests on water-shedding sites. A nitrophytic species characteristic of Moder and Mull humus forms.

Saxifraga ferruginea
Alaska saxifrage

Saxifragaceae
(Saxifrage family)

A shade-intolerant, submontane to alpine, Western North American forb distributed more in the Pacific than the Cordilleran region. Occurs in water-shedding sites within alpine, tundra, boreal, cool temperate, and cool mesothermal climates. Common but scattered in non-forested communities and open-canopy forests. Usually inhabits very shallow soils on rocks subjected to severe desiccation but often affected by temporary surface seepage. Frequently associated with *Pleurozium schreberi*, *Rhacomitrium canescens*, and lichens. Characteristic of early-seral communities.

Saxifraga tolmiei
Tolmie's saxifrage

Saxifragaceae
(Saxifrage family)

A very shade-intolerant, subalpine to alpine, Western North American forb distributed more in the Pacific than the Cordilleran region. Occurs in maritime to submaritime alpine tundra and subalpine boreal climates on very moist to wet exposed mineral soils. Common in sparsely vegetated, non-forested snow-patch communities on unstable sheet-wash slopes or hamada-like surfaces; often associated with graminoids. Characteristic of alpine communities.

Scapania bolanderi

Scapaniaceae

A shade-tolerant, submontane to subalpine, Asian and Western North American liverwort distributed more in the Pacific than the Cordilleran region. Occurs in cool temperate and cool mesothermal climates on nitrogen-poor soils; its occurrence decreases with increasing continentality. Plentiful on decaying coniferous wood in coniferous forests, common on the forest floor in Queen Charlotte Islands. Usually associated with *Lepidozia reptans*, *Plagiothecium undulatum*, *Rhizomnium glabrescens*, and *Rhytidiadelphus loreus*. An oxylophytic species characteristic of Mor humus forms.

Scirpus microcarpus
Small-flowered bulrush

Cyperaceae
(Sedge family)

A shade-intolerant, submontane to subalpine, Asian and Western North American graminoid distributed equally in the Pacific and Cordilleran regions and marginally in the Central region. Occurs in cool temperate and cool mesothermal climates on wet to very wet, nitrogen-rich soils (Moder and Mull humus forms); its occurrence increases with increasing precipitation and decreases with elevation. Common and often dominant in semiterrestrial, usually early-seral, graminoid-dominated communities and disturbed sites. Often associated with *Carex sitchensis*, occasionally associated with *Lysichitum americanum*. A nitrophytic species characteristic of nutrient-rich wetlands.

Sedum spathulifolium
Broad-leaved stonecrop

Crassulaceae
(Stonecrop family)

A very shade-intolerant, submontane to subalpine, Western North American forb distributed more in the Pacific than the Cordilleran region. Occurs in maritime to submaritime summer-dry cool mesothermal climates on excessively dry to very dry, nitrogen-poor soils; its occurrence decreases with increasing precipitation and continentality. Sparse to sporadic in non-forested communities on very shallow, water-shedding sites (rock outcrops and cliffs). Characteristic of moisture-deficient sites.

Selaginella wallacei
Wallace's selaginella

Selaginellaceae

A shade-intolerant, submontane to subalpine, Western North American fern-ally distributed equally in the Pacific and Cordilleran regions. Occurs on very shallow, very dry to moderately dry, nitrogen-poor soils within boreal, temperate, cool semiarid, and cool mesothermal climates. Its occurrence decreases with increasing latitude, elevation and precipitation. Commonly scattered in non-forested communities on very shallow soils on rock outcrops. Usually associated with *Arctostaphylos uva-ursi, Cryptogramma crispa, Rhacomitrium canescens,* and lichens. Characteristic of moisture-deficient sites.

Senecio vulgaris
Common groundsel
Senecio sylvaticus
Wood groundsel

Asteraceae
(Aster family)

Shade-intolerant, submontane to subalpine, European forbs introduced to Pacific and Atlantic North America (*S. sylvaticus*) or transcontinental (*S. vulgaris*). Occur in maritime to submaritime climates on very moist to wet, nitrogen-rich soils; their occurrence decreases with increasing elevation and continentality.

Both species are scattered to plentiful, occasionally dominant, in non-forested communities on cutover, fire-disturbed, or continuously disturbed sites, where they inhabit exposed mineral soils. Often associated with *Anaphalis margaritacea, Hieracium albiflorum, Hypochaeris radicata, Holcus lanatus,* and *Mycelis muralis.* Nitrophytic species characteristic of early-seral communities.

Senecio triangularis
Arrow-leaved groundsel

Asteraceae
(Aster family)

A shade-tolerant to very shade-intolerant, montane to alpine, Western North American forb distributed equally in the Pacific and Cordilleran regions. Occurs on very moist to wet, nitrogen-rich soils within alpine tundra, boreal, cool temperate, and cool mesothermal climates; its occurrence increases with precipitation. Scattered to plentiful in meadow-like communities and the herbaceous understory of open-canopy forests on water-receiving and water-collecting sites. A nitrophytic species characteristic of Moder and Mull humus forms.

Shepherdia canadensis
Soopolallie

Elaeagnaceae
(Oleaster family)

A shade-tolerant/intolerant, submontane to subalpine, transcontinental North American deciduous shrub (sporadic in the Pacific region). Occurs predominantly in continental boreal and cool temperate climates on very dry to moderately dry, nitrogen-medium soils; its occurrence increases with increasing continentality. Common in semi-open forests on water-shedding sites; scattered on the leeward side of Vancouver Island; plentiful in the coast-interior ecotone. Often associated with *Calamagrostis rubescens*, *Linnaea borealis*, and *Paxistima myrsinites*. Symbiotic with nitrogen-fixing organisms. Characteristic of continental forests.

217

Sibbaldia procumbens
Sibbaldia

Rosaceae
(Rose family)

A very shade-intolerant, sub-alpine to alpine, circumpolar evergreen dwarf shrub (trans-continental in North America). Occurs in alpine tundra and subalpine boreal climates on fresh to moist, nitrogen-poor soils (Mor humus forms); its occurrence increases with increasing latitude and duration of snow cover. Sporadic to scattered in heath communities and parkland forests on water-shedding sites. Usually associated with *Barbilophozia floerkei* and ericaceous shrubs. An oxylophytic species characteristic of alpine and subalpine communities.

Siphula ceratites

Usneaceae

A shade-intolerant, montane to alpine, circumpolar lichen distributed equally in the Pacific and Atlantic regions. Occurs on wet to very wet, nitrogen-poor soils (Mor humus forms) within hypermaritime to maritime alpine tundra boreal, and summer-wet cool mesothermal climates. Sporadic to scattered on topographic prominences in semi-terrestrial communities and open-canopy subalpine forests on water-collecting sites. Often associated with *Sphagnum* species. An oxylophytic species characteristic of nutrient-poor wetlands.

Sisyrinchium douglasii
Satin-flower

Iridaceae
(Iris family)

A shade-intolerant, submontane to subalpine, Western North American forb distributed more in the Pacific than the Cordilleran region. Occurs in maritime to submaritime summer-dry mesothermal climates on very dry to moderately dry, nitrogen-medium soils; its occurrence decreases with increasing latitude, elevation, and precipitation. Sporadic in meadow-like communities and open-canopy forests on base-rich, water-shedding sites (rock outcrops, cliffs, and colluvial slopes) affected by temporary seepage in spring. Characteristic of moisture-deficient sites.

Smilacina stellata
Star-flowered false Solomon's-sea
Smilacina racemosa
False Solomon's-seal

Liliaceae
(Lily family)

S. stellata - a shade-tolerant to very shade-intolerant, submontane to subalpine, transcontinental North American forb. Occurs in boreal, temperate, and mesothermal climates on nitrogen-rich soils. *S. racemosa* - a shade-tolerant to shade-intolerant, submontane to subalpine, transcontinental North American forb. Occurs in cool temperate and cool mesothermal climates on nitrogen-rich soils.

Widespread, scattered to plentiful on water-shedding sites (south-facing colluvial slopes); plentiful on water-receiving sites (*S. stellata* common on floodplains); persist on clearings; their occurrence decreases with increasing latitude and elevation. Often associated with *Acer glabrum*, *Disporum hookeri*, *Osmorhiza chilensis*, *Polystichum munitum*, *Populus trichocarpa*, and *Tiarella trifoliata*. Nitrophytic species characteristic of Moder and Mull humus forms.

Sorbus sitchensis
Sitka mountain-ash
Sorbus scopulina
Western mountain-ash

Rosaceae
(Rose family)

S. sitchensis - a shade-intolerant, montane to subalpine, Western North American deciduous shrub distributed more in the Pacific than the Cordilleran region. Occurs in maritime to submaritime subalpine boreal climates on moderately dry to fresh, nitrogen-poor soils; its occurrence increases with increasing precipitation and elevation, and decreases with continentality. *S. scopulina* - a shade-tolerant/intolerant to very shade-intolerant, montane to subalpine, Western North American deciduous shrub distributed in the Cordilleran region, marginally in the Pacific and Central regions. Occurs in continental boreal and wet cool temperate climates on moderately dry to fresh, nitrogen-medium soils; its occurrence increases with increasing continentality.

Both species are common but scattered in open-canopy, coniferous forests on watershedding sites; persist in clearings. Characteristic of Mor humus forms.

Sphagnum fuscum
Sphagnum rubellum

Sphagnaceae

Shade-intolerant, mostly circumpolar mosses, transcontinental in North America, occuring from cool mesothermal climates to boreal climates. Peat-forming species of open semi-terrestrial communities with high water table, wet to very wet nitrogen-poor organic soils. Often associated with *Vaccinium oxycoccus, Ledum groenlandicum, Coptis trifolia Trientalis artcica.* Oxylophytic species characteristic of nutrient-poor organic soils.

Several other species of this genus, such as *S. capillifolium* (= *S. nemoreum*), *S. fallax, S. papillosum* and *S. tenellum* have similar ecological requirements and all indicate nutrient-poor, acidic organic soils.

Sphagnum girgensohnii   *Sphagnaceae*

A shade-tolerant, submontane to subalpine circumpolar moss, transcontinental in North America. Occurs in cool meso-thermal climates on very moist to wet soils in coniferous forest on receiving sites with gleysolic or organic soils. Often associated with *Blechnum spicant* and *Coptis aspleniifolia*.

Spiraea betulifolia   *Rosaceae*
Birch-leaved spirea   (Rose family)

A shade-tolerant/intolerant, submontane to subalpine, Asian and Cordilleran North American (marginally Central) deciduous shrub. Occurs in continental boreal and cool temperate climates on very dry to moderately dry, nitrogen-medium soils; its occurrence increases with increasing continentality and decreases with increasing latitude and precipitation. Common, often plentiful, in semi-open forests (persists on cutover areas) on water-shedding sites in the eastern part of the coast-interior ecotone. Usually associated with *Acer glabrum*, *Amelanchier alnifolia*, *Aster conspicuus*, *Calamagrostis rubescens*, *Mahonia aquifolium*, *Rhytidiadelphus triquetrus*, and *Shepherdia canadensis*. Characteristic of continental forests.

Spiraea densiflora
Subalpine spirea

Rosaceae
(Rose family)

A shade-intolerant, montane to alpine, Western North American deciduous shrub distributed more in the Pacific than the Cordilleran region. Occurs in continental subalpine boreal and cool temperate climates on fresh to moist, nitrogen-medium soils; its occurrence increases with increasing latitude and continentality. Sporadic to scattered in meadow-like communities and open-canopy forests on water-shedding and water-receiving sites in the eastern coast-interior ecotone. Characteristic of continental forests.

Spiraea douglasii
Hardhack

Rosaceae
(Rose family)

A shade-intolerant, submontane to subalpine, Western North American deciduous shrub distributed more in the Pacific than the Cordilleran region. Occurs in cool mesothermal climates on very moist to wet, nitrogen-medium soils; its occurrence decreases with increasing latitude, elevation, and continentality. Common, occasionally dominant, in semi-terrestrial communities and open-canopy forests on disturbed water-receiving and water-collecting sites. Tolerates fluctuating groundwater tables. Often associated with *Gaultheria shallon*, *Juncus effusus*, *Myrica gale*, or *R.spectabilis*. Characteristic of wetlands.

Spiraea menziesii
Spiraea douglasii ssp. *menziesii*
Pink spirea

Rosaceae
(Rose family)

A shade-intolerant, submontane to subalpine, Western
North American deciduous
shrub distributed more in the
Cordilleran than the Pacific
region. Occurs on very moist to
wet, nitrogen-rich soils; its
occurrence increases with increasing continentality and
decreases with increasing latitude. Sporadic in semi-terrestrial communities and open-
canopy forests in boreal, cool
temperate, and cool mesothermal climates on water-receiving (floodplain) and water-collecting sites. Tolerates fluctuating groundwater tables.
Characteristic of nutrient-rich
sites.

Stachys cooleyae
Cooley's hedge-nettle
Stachys mexicana
Mexican hedge-nettle

Lamiaceae
(Mint family)

Shade-tolerant/intolerant, submontane to
montane, Pacific North American forbs. Both
species occur in cool mesothermal climates on
very moist to wet, nitrogen-rich soils. Their
occurrence decreases with increasing latitude, elevation, and continentality. Sporadic
in the herbaceous understory of broad-leaved
forests on water-receiving (floodplain) sites.
Commonly associated with *Adiantum pedatum,*
Athyrium filix-femina, Circaea pacifica, Oplo-
panax horridus, Tiarella trifoliata, and *Tolmiea*
menziesii. Nitrophytic species characteristic
of Moder and Mull humus forms.

Stellaria calycantha
Northern starwort

Caryophyllaceae
(Pink family)

A shade-tolerant/intolerant, montane to subalpine, circumpolar forb (transcontinental in North America). Occurs in continental boreal and cool temperate climates on very moist to wet soils; its occurrence increases with increasing latitude, elevation, and continentality. Sparse to sporadic on water-receiving and water-collecting sites in the coast-interior ecotone. Characteristic of continental forests.

Stellaria crispa
Crisp starwort

Caryophyllaceae
(Pink family)

A shade-tolerant/intolerant, submontane to montane, Western North American forb distributed more in the Pacific than the Cordilleran region. Occurs on very moist to wet, nitrogen-rich soils within boreal, temperate, and cool mesothermal climates; its occurrence increases with increasing latitude and decreases with increasing elevation and continentality. Sporadic to plentiful in broad-leaved forests on water-receiving and water-collecting sites; occasionally inhabits disturbed sites. A nitrophytic species characteristic of Moder and Mull humus forms.

Stenanthium occidentale
Mountainbells

Liliaceae
(Lily family)

A shade-intolerant, subalpine to alpine, Western North American forb distributed more in the Pacific than the Cordilleran region. Occurs in subalpine boreal climates on fresh to very moist, nitrogen-rich soils (Moder and Mull humus forms); its occurrence increases with increasing elevation and decreases with continentality. Sporadic in non-forested communities and open-canopy forests on water-receiving, late snow-melt sites. Characteristic of subalpine forests.

Stereocaulon tomentosum

Stereocaulaceae

A shade-intolerant, submontane to subalpine, South American and circumpolar lichen (transcontinental in North America). Contains nitrogen-fixing blue-green algae. Occurs on excessively dry to very dry, nitrogen-poor soils (Mor humus forms) within tundra, boreal, temperate, and cool mesothermal climates. Sporadic to plentiful in non-forested communities and open-canopy forests on water-shedding sites. Inhabits substrates subjected to rapid desiccation, usually associated with lichens. An oxylophytic species characteristic of moisture-deficient sites.

Streptopus amplexifolius
Clasping twistedstalk
Streptopus roseus
Rosy twistedstalk

Liliaceae
(Lily family)

S. amplexifolius - a shade-tolerant/intolerant, submontane to subalpine, circumpolar forb (transcontinental in North America). *S. roseus* - a shade-tolerant/intolerant, montane to subalpine, transcontinental North American forb. Both species occur on fresh to very moist, nitrogen-rich soils within boreal, temperate, and cool mesothermal climates. Their occurrence increases with increasing latitude and precipitation. Sporadic to abundant in submontane to subalpine coniferous forests on water-receiving and water-collecting sites. Usually associated with *Abies amabilis. A. lasiocarpa, Athyrium filix-femina, Gymnocarpium dryopteris, Oplopanax horridus, Ribes lacustre, Rubus parviflorus, R. spectabilis, Tiarella trifoliata*, and *T. unifoliata*. Nitrophytic species characteristic of Moder and Mull humus forms.

Streptopus streptopoides
Small twistedstalk

Liliaceae
(Lily family)

A shade-tolerant/intolerant, montane to subalpine, Asian and North American forb distributed more in the Pacific than the Cordilleran region. Occurs in subalpine boreal and cool temperate climates on fresh to very moist, nitrogen-poor soils; its occurrence decreases with increasing continentality. Sporadic to scattered in mature to subalpine coniferous forests on water-shedding and water-receiving sites. Typically associated with *Blechnum spicant, Dryopteris expansa, Orthilia secunda, Rhytidiopsis robusta, Vaccinium alaskaense*, and *V. membranaceum*. An oxylophytic species characteristic of Mor humus forms.

Symphoricarpos albus
Common snowberry

Caprifoliaceae
(Honeysuckle family)

A shade-tolerant/intolerant, submontane to montane, transcontinental North American deciduous shrub. Occurs on nitrogen-rich soils within boreal, temperate, cool semiarid, and mesothermal climates; its occurrence decreases with increasing elevation and precipitation. Tolerates fluctuating ground-water tables. Scattered in coniferous forests, plentiful in broad-leaved forests on water-shedding and water-receiving (floodplain) sites; persists on cutover areas. Commonly associated with *Aster conspicuus, Calamagrostis rubescens, Mahonia aquifolium, M. nervosa, Spiraea betulifolia,* and *Rhytidiadelphus triquetrus.* Characteristic of Moder and Mull humus forms.

Symphoricarpos hesperius
Symphoricarpos mollis ssp. *hesperius*
Trailing snowberry

Caprifoliaceae
(Honeysuckle family)

A shade-intolerant, submontane to subalpine, Western North American shrub distributed more in the Pacific than the Cordilleran region. Occurs in maritime to submaritime summer-dry cool mesothermal climates on very dry to moderately dry, nitrogen-medium soils. Its occurrence decreases with increasing latitude, elevation, precipitation, and continentality. Sporadic in open-canopy Douglas-fir forests on water-shedding sites. Often associated with *Campanula scouleri, Lonicera ciliosa, L. hispidula, Mahonia aquifolium, M. nervosa,* and *Rhytidiadelphus triquetrus.* Characteristic of moisture-deficient sites.

Taxus brevifolia
Pacific yew, western yew

Taxaceae
(Yew family)

A shade-tolerant, submontane to subalpine, Western North American conifer distributed equally in the Pacific and Cordilleran regions. Occurs in cool temperate and cool mesothermal climates; its occurrence increases with increasing precipitation and decreases with increasing latitude and elevation. Scattered in coniferous forests; common on water-receiving sites, frequent on water-collecting sites, and occasional on water-shedding sites. Characteristic of temperate and mesothermal coniferous forests.

Tellima grandiflora
Tall fringecup

Saxifragaceae
(Saxifrage family)

A shade-tolerant/intolerant, submontane to montane, Western North American forb distributed more in the Pacific than the Cordilleran region. Occurs in cool temperate and cool mesothermal climates on fresh to very moist, nitrogen-rich soils; its occurrence decreases with increasing continentality and elevation. Frequent on water-receiving sites and sporadic on water-shedding sites; most common in broad-leaved forests on floodplains where it associates with *Acer macrophyllum, Athyrium filix-femina, Polystichum munitum, Sambucus racemosa, Tiarella trifoliata,* and *Tolmiea menziesii.* A nitrophytic species characteristic of Moder and Mull humus forms.

Thalictrum occidentale
Western meadowrue

Ranunculaceae
(Buttercup family)

A shade-tolerant/intolerant, submontane to subalpine, Western North American forb distributed equally in the Pacific and Cordilleran regions, and marginally in the Central region. Occurs on fresh to very moist, nitrogen-rich soils within montane boreal, temperate, and cool mesothermal climates; its occurrence decreases with increasing elevation and latitude. Common and often plentiful in the herbaceous understory of open-canopy forests on water-shedding and water-receiving sites; occasionally inhabits exposed mineral soils on disturbed sites; tolerates fluctuating groundwater tables. Often associated with *Athyrium filix-femina*, *Cornus sericea*, *Gymnocarpium dryopteris*, *Lonicera involucrata*, *Oplopanax horridus*, *Ribes lacustre*, and *Tiarella trifoliata*. A nitrophytic species characteristic of Moder and Mull humus forms.

Thuja plicata
Western redcedar

Cupressaceae
(Cypress family)

A shade-tolerant to shade-tolerant/intolerant, submontane to subalpine, Western North American evergreen conifer distributed more in the Pacific than the Cordilleran region. Occurs predominantly in cool temperate and cool mesothermal climates; its occurrence decreases with increasing latitude, elevation, and continentality. One of the most common trees in central and southern B.C.; often forms pure stands on floodplains and wetland sites. As does yellow-cedar, western redcedar tolerates a nearly complete edaphic range, and develops a very dense root system; the latter feature may explain its abundance on very steep, seepage-affected, and often unstable colluvial soils. Most productive on submontane, fresh to moist, nutrient-rich soils within cool mesothermal climates. Characteristic of cool temperate and mesothermal forests.

Tiarella trifoliata
Three-leaved foam-flower
Tiarella laciniata
Cut-leaved foam-flower

Saxifragaceae
(Saxifrage family)

Shade-tolerant/intolerant, submontane to subalpine, Western North American forbs distributed more in the Pacific than in the Cordilleran region. These species occur in hypermaritime to maritime cool mesothermal (*T. laciniata*) climate on fresh to very moist, nitrogen-rich soils; their occurrence decreases with increasing latitude, elevation, and continentality. Scattered to abundant on water-receiving sites that support a very productive growth of Douglas-fir, Sitka spruce, true firs, and western redcedar. Often grow with *Achlys triphylla*, *Athyrium filix-femina*, *Galium triflorum*, *Polystichum munitum*, *Rubus parviflorus*, and *R. spectabilis*. Nitrophytic species characteristic of Moder and Mull humus forms.

Tiarella unifoliata
One-leaved foam-flower

Saxifragaceae
(Saxifrage family)

A shade-tolerant/intolerant, submontane to subalpine, Western North American forb distributed equally in the Pacific and Cordilleran regions. Occurs in boreal and cool temperate climates on fresh to very moist, nutrient-medium soils; its occurrence increases with increasing elevation, continentality, and precipitation. Occasional on water-shedding sites, plentiful on water-receiving sites. Commonly associated with *Athyrium filix-femina*, *Gymnocarpium dryopteris*, *Oplopanax horridus*, *Ribes lacustre*, and *Streptopus roseus*; often grows with *Clintonia uniflora*, *Paxistima myrsinites*, *Rhytidiopsis robusta*, or *Vaccinium membranaceum*. Characteristic of Mor and acidic Moder humus forms.

Timmia austriaca

Timmiaceae

A shade-tolerant/intolerant, submontane to alpine, circumpolar moss (transcontinental in North America). Occurs on nitrogen-medium soils within tundra, boreal, cool temperate, and cool mesothermal climates; its occurrence increases with increasing elevation. Very sparse in coastal forests on water-shedding and water-receiving sites; usually inhabits exposed mineral soil or friable organic materials on colluvial slopes. Characteristic of calcium-rich substrates.

Tofieldia glutinosa
Sticky false asphodel

Liliaceae
(Lily family)

A shade-intolerant, submontane to alpine, transcontinental North American forb. Occurs on wet to very wet, calcium-rich soils within boreal, temperate, and cool mesothermal climates; its occurrence increases with increasing precipitation. Sporadic in non-forested, semi-terrestrial communities on water-collecting sites. Characteristic of wetlands.

Tolmiea menziesii
Piggy-back plant, youth-on-age

Saxifragaceae
(Saxifrage family)

A shade-tolerant/intolerant, submontane to subalpine, Pacific North American forb (very marginally in the Cordilleran region). Occurs in maritime to submaritime cool mesothermal climates on fresh to very moist, nitrogen-rich soils; its occurrence decreases with latitude, elevation, and continentality. Sporadic in coniferous forests, plentiful in broad-leaved forests on water-receiving sites. Associated with *Athyrium filix-femina*, *Galium triflorum*, *Oplopanax horridus*, *Polystichum munitum*, *Rubus spectabilis*, and *Tiarella trifoliata*. A nitrophytic species characteristic of Moder and Mull humus forms.

Torreyochloa pauciflora
Puccinellia pauciflora
Glyceria pauciflora
Weak false-manna

Poaceae
(Grass family)

A shade-tolerant/intolerant, submontane to alpine, Western North American grass distributed equally in the Pacific and Cordilleran regions, and marginally in the Central region. Occurs on wet to very wet, nitrogen-rich soils (Moder and Mull humus forms) within boreal, temperate, and cool mesothermal climates; its occurrence decreases with increasing elevation. Sporadic in non-forested, semi-terrestrial communities on water-collecting sites, often along lakes; associated with graminoids. A nitrophytic species characteristic of nutrient-rich wetlands.

Trautvetteria caroliniensis
False bugbane

Ranunculaceae
(Buttercup family)

A shade-tolerant/intolerant, submontane to montane, Asian and transcontinental North American forb (absent in the Central region). Occurs in cool temperate and cool mesothermal climates on fresh to very moist, nitrogen-rich soils; its occurrence decreases with increasing latitude, elevation and continentality. Scattered, occasionally dominant, in broad-leaved forests on water-receiving sites (alluvium, floodplain, and stream-edge); extends to water-collecting sites. Tolerates fluctuating groundwater tables. Often associated with *Athyrium filix-femina*, *Boykinia elata*, *Maianthemum dilatatum*, *Polystichum munitum*, *Rubus spectabilis*, and *Tiarella trifoliata*. A nitrophytic species characteristic of Moder and Mull humus forms.

Trientalis arctica
Northern starflower
Trichophorum cespitosum
Baeothryon cespitosum, Scirpus cespitosus
Tufted deergrass, tufted clubrush

Primulaceae
(Primerose family)
Cyperaceae
(Sedge family)

Tricentalis arctica - a shade-intolerant, submontane to subalpine, circumpolar forb distributed equally in Pacific, Cordilleran, and (marginally) Central North America. *Trichophorum cespitosum* - a very shade-intolerant, submontane to subalpine, circum-polar sedge (transcontinental in North America).

These species occur predominantly on wet to very wet, nitrogen-poor soils (Mor humus forms) in boreal and cool temperate climates; their occurrence increases with increasing latitude. Sporadic in non-forested, semi-terrestrial communities on water-collecting sites; usually associated with *Sphagnum* species. Oxylophytic species characteristic of nutrient-poor wetlands.

Trientalis latifolia
Broad-leaved starflower

Primulaceae
(Primrose family)

A shade-tolerant/intolerant, submontane to subalpine, Western North American forb distributed equally in the Pacific and Cordilleran regions. Occurs in cool temperate and cool mesothermal climates on moderately dry to fresh, nitrogen-medium soils; its occurrence decreases with increasing elevation and latitude. Scattered to plentiful on water-shedding sites, less frequent on water-receiving sites. Usually associated with *Acer glabrum, Kindbergia oregana, Mahonia nervosa,* and *Polystichum munitum.* Characteristic of young-seral forests.

Trillium ovatum
Western trillium

Liliaceae
(Lily family)

A shade-tolerant/intolerant, submontane to montane, Western North American forb distributed more in the Pacific than the Cordilleran region. Occurs in maritime to submaritime cool mesothermal climates on fresh to very moist, nitrogen-rich soils; its occurrence decreases with increasing elevation, latitude, and continentality. Scattered on water-receiving sites; usually associated with *Achlys triphylla, Galium triflorum, Polystichum munitum, Streptopus amplexifolius,* and *Tiarella trifoliata.* A nitrophytic species characteristic of Moder and Mull humus forms.

Trisetum cernuum
Nodding trisetum

Poaceae
(Grass family)

A shade-tolerant/intolerant, submontane to subalpine, Western North American grass distributed more in the Pacific than the Cordilleran region. Occurs on fresh to very moist, nitrogen-rich soils within boreal, wet temperate, and cool mesothermal climates; its occurrence increases with increasing precipitation and decreases with increasing continentality. Scattered in coniferous forests, plentiful in broad-leaved forests on water-receiving (alluvial, floodplain, and stream-edge) sites; often inhabits exposed mineral soils. Frequently associated with *Elymus hirsutus, Galium triflorum, Polystichum munitum, Ribes lacustre, Rubus parviflorus,* and *Tiarella trifoliata.* A nitrophytic species characteristic of Moder and Mull humus forms.

Tsuga heterophylla
Western hemlock

Pinaceae
(Pine family)

A very shade-tolerant, submontane to subalpine, Western North American evergreen conifer distributed more in the Pacific than the Cordilleran region. Occurs in cool temperate and cool mesothermal climates; its occurrence increases with precipitation, and decreases with increasing elevation and continentality. Most productive on submontane, fresh, and nutrient-medium soils within summer-wet cool mesothermal climates.

Common on nitrogen-poor, water-shedding and water-receiving sites; on nitrogen-rich soils restricted to acid organic materials. Regenerate on acid organic substrates and on decaying coniferous wood.

Tsuga mertensiana
Mountain hemlock

(Pinaceae)
(Pine family)

A shade-tolerant/intolerant, montane to alpine, Western North American evergreen conifer distributed more in the Pacific than the Cordilleran region. Occurs in hypermaritime to submaritime subalpine boreal climates; its occurrence decreases with increasing precipitation and continentality. Most productive on fresh and nutrient-medium soils within maritime subalpine boreal climates. Characteristic of maritime subalpine forests.

Urtica lyallii
Urtica dioica ssp. *gracilis* var. *lyallii*
Stinging nettle

Urticaceae
(Nettle family)

A shade-tolerant/intolerant, submontane to subalpine, circumpolar forb (transcontinental in North America). Occurs on fresh to very moist, nitrogen-rich soils (Moder and Mull humus forms) within boreal, temperate, cool semiarid, and mesothermal climates; its occurrence decreases with increasing elevation. Frequent in herbaceous communities, occasional in broad-leaved forests on disturbed, water-shedding and water-receiving sites. A nitrophytic species characteristic of continuously disturbed sites.

Usnea longissima   *Usneaceae*

A shade-intolerant, submontane to subalpine, circumpolar epiphytic lichen; distributed in Pacific, Cordilleran (less frequently), and Atlantic North America. Occurs in summer-wet cool mesothermal climates; its occurrence decreases with increasing elevation and continentality. Scattered to plentiful on trunks and branches of conifers in open-canopy forests on water-receiving (floodplain) sites. Characteristic of wet mesothermal forests.

Vaccinium alaskaense   *Ericaceae*
Alaskan blueberry   (Heath family)

A shade-tolerant, submontane to subalpine, Pacific North American deciduous shrub. Occurs in hypermaritime to maritime subalpine boreal and summer-wet cool mesothermal climates on fresh to very moist, nitrogen-poor soils; its occurrence decreases with increasing precipitation and continentality.

Vaccinium caespitosum
Dwarf blueberry

Ericaceae
(Heath family)

A shade-tolerant/intolerant, submontane to alpine, transcontinental North American deciduous shrub. Occurs in boreal and cool temperate climates on fresh to very moist, nitrogen-poor soils; its occurrence increases with increasing latitude and continentality.

Vaccinium deliciosum
Cascade blueberry, blue-leaved blueberry

Ericaceae
(Heath family)

A shade-intolerant, submontane to subalpine, Western North American deciduous shrub distributed more in the Pacific than the Cordilleran region. Occurs in maritime to submaritime subalpine boreal climates on fresh to very moist, nitrogen-poor soils; its occurrence increases with increasing elevation and decreases with increasing continentality. Inhabits late snow-melt sites.

Vaccinium membranaceum
Black huckleberry

Ericaceae
(Heath family)

A shade-tolerant/intolerant, montane to subalpine, Western North American deciduous shrub distributed more in the Cordilleran than the Pacific region (reported in the Central region). Occurs in boreal and cool temperate climates on moderately dry to fresh, nitrogen-poor soils; its occurrence increases with increasing elevation and continentality.

Vaccinium myrtilloides
Velvet-leaved blueberry

Ericaceae
(Heath family)

A shade-tolerant/intolerant, submontane to subalpine, transcontinental North American deciduous shrub (rare in the Pacific region, sporadic in the eastern limits of the coast-interior ecotone). Occurs in continental boreal and cool temperate climates on nitrogen-poor soils; its occurrence increases with increasing continentality.

Vaccinium ovalifolium
Oval-leaved blueberry

Ericaceae
(Heath family)

A shade-tolerant, submontane to subalpine, Asian and transcontinental North American deciduous shrub mostly in the Pacific and Cordilleran regions, rare in the Central and Atlantic regions. Occurs on fresh to very moist, nitrogen-poor soils with boreal, cool temperate, and cool mesothermal climates; its occurrence increases with increasing latitude and decreases with increasing continentality.

Vaccinium ovatum
Evergreen huckleberry

Ericaceae
(Heath family)

A shade-intolerant, submontane to montane, Pacific North American evergreen shrub. Occurs in hypermaritime to maritime summer-wet cool mesothermal climates on moderately dry to fresh, nitrogen-poor soils; its occurrence decreases with increasing elevation and increases with increasing precipitation.

Vaccinium oxycoccos
Oxycoccus palustris
Bog cranberry

Ericaceae
(Heath family)

A shade-intolerant, submontane to montane, circumpolar evergreen shrub (transcontinental in North America). Occurs on wet to very wet, nitrogen-poor soils within montane boreal, cool temperate, and cool mesothermal climates in nutrient-poor wetlands; its occurrence increases with increasing latitude and continentality.

Vaccinium parvifolium
Red huckleberry

Ericaceae
(Heath family)

A shade-tolerant, submontane to subalpine, Western North American deciduous shrub distributed more in the Pacific than the Cordilleran region. Occurs in cool mesothermal climates on nitrogen-poor soils; its occurrence decreases with increasing elevation and continentality.

Vaccinium scoparium
Grouseberry

Ericaceae
(Heath family)

A shade-tolerant/intolerant, montane to alpine, Cordilleran North American deciduous shrub (marginal in the Central region). Occurs in continental alpine tundra and subalpine boreal climates on nitrogen-poor soils; its occurrence is restricted to the eastern part of the coast-interior ecotone.

Vaccinium uliginosum
Bog blueberry

Ericaceae
(Heath family)

A shade-intolerant, submontane to alpine, circumpolar deciduous shrub (transcontinental in North America). Occurs on very moist to wet, nitrogen-poor soils within tundra, boreal, cool temperate, and cool mesothermal climates.

Vahlodea atropurpurea
Deschampsia atropurpurea
Mountain hairgrass

Poaceae
(Grass family)

A shade-intolerant, subalpine to alpine, cir-cumpolar grass (transcontinental in North America). Occurs in alpine tundra and sub-alpine boreal climates on fresh to very moist, nitrogen-medium soils. Sporadic in non-for-ested, heath or graminoid-forb communities on water-shedding and water-receiving sites (springs, streams, and snow basins). Often as-sociated with *Cassiope mertensiana*, *Leptarrhena pyrolifolia*, *Phyllodoce empetri-formis*, and *Vaccinium deliciosum*. Character-istic of alpine and subalpine communities.

Valeriana sitchensis
Sitka valerian
Valeriana scouleri
Valeriana sitchensis ssp. *scouleri*
Scouler's valerian

Valeriaceae
(Valaerian family)

Shade-tolerant/intolerant, submontane to alpine, West-ern North American forbs dis-tributed equally in the Pacific and Cordilleran regions. Oc-cur on fresh to very moist, ni-trogen-rich soils within boreal, cool temperate, and cool meso-thermal climates. Sporadic (*V. scouleri*) to common (*V. sitch-ensis*) in meadow-like commu-nities and coniferous forests on water-receiving and water-collecting sites. Usually asso-ciated with *Athyrium filix-femina*, *Gymnocarpium dry-opteris*, *Lonicera utahensis*, *Oplopanax horridus*, and *Streptopus roseus*. Nitrophytic species characteristic of Moder and Mull humus forms.

Veratrum eschscholtzii

Veratrum viride ssp. *eschscholtzii*
Indian false hellebore, American False hellebore, Indian-poke

Liliaceae
(Lily family)

A shade-intolerant, submontane to alpine, transcontinental North American forb (rare in the Central region). Occurs on very moist to wet, nitrogen-rich soils within alpine tundra, boreal, cool temperate, and cool mesothermal climates. Common in non-forested, semi-terrestrial communities and open-canopy subalpine forests on water-receiving and water-collecting sites. Usually associated with *Athyrium filix-femina, Coptis aspleniifolia, Lysichitum americanum, Oplopanax horridus*, and *Streptopus roseus*. A nitrophytic species characteristic of Moder and Mull humus forms.

Veronica americana
American brooklime

Scrophulariaceae
(Figwort family)

A shade-intolerant, submontane to subalpine, Asian and transcontinental North American forb. Occurs on fresh to very moist, nitrogen-rich soils (Moder and Mull humus forms) within boreal, temperate, and mesothermal climates. Scattered in herbaceous, non-forested communities and open-canopy forests on water-receiving (floodplains and stream-edge) sites. A nitrophytic species characteristic of alluvial floodplain forests.

Viburnum edule
Highbush-cranberry
Viburnum trilobum
Viburnum opulus
American bush-cranberry

Caprifoliaceae
(Honeysuckle family)

Shade-tolerant/intolerant, submontane to subalpine, transcontinental North American deciduous shrubs. Occur in continental boreal and cool temperate climates (*V. edule*) or in cool temperate and cool mesothermal climates (*V. trilobum*) on fresh to very moist, nitrogen-rich soils; their occurrence increases with increasing continentality. Rare to common (especially in broad-leaved forests) on water-receiving and water-collecting sites. Nitrophytic species characteristic of Moder and Mull humus forms.

Vicia americana
American vetch

Fabaceae
(Pea family)

A shade-tolerant/intolerant, submontane to subalpine, transcontinental North American forb. Occurs on moderately dry to fresh, nitrogen-medium soils within boreal, temperate, and mesothermal climates; its occurrence decreases with elevation. Scattered in young-seral forests on water-shedding and water-receiving sites. Symbiotic with nitrogen-fixing organisms.

Viola adunca
Early blue violet

Violaceae
(Violet family)

A shade-intolerant, submontane to alpine, transcontinental North American forb. Occurs on very dry to moderately dry, nitrogen-medium soils within boreal, temperate, and mesothermal climates; its occurrence decreases with elevation and precipitation.

Viola glabella
Stream violet

Violaceae
(Violet family)

A shade-tolerant to very shade-intolerant, submontane to subalpine, Western North American herb distributed equally in the Pacific and Cordilleran regions. Occurs on very moist to wet, nitrogen-rich soils (Moder and Mull humus forms) within boreal, temperate, and cool mesothermal climates; its occurrence increases with increasing latitude and precipitation. Common in non-forested communities and open-canopy forests on water-receiving (flood-plains and stream-edge) sites. Usually associated with *Athyrium filix-femina*, *Cornus sericea*, *Lysichitum americanum*, *Rubus spectabilis*, and *Tiarella trifoliata*. A nitrophytic species characteristic of flooded sites.

Viola orbiculata
Round-leaved violet
Viola sempervirens
Trailing yellow violet

Violaceae
(Violet family)

V. orbiculata - a shade-tolerant/ intolerant, montane to alpine, Western North American forb distributed equally in the Pacific and Cordilleran regions. Occurs on moderately dry to fresh, nitrogen-medium soils within boreal, wet cool temperate, and cool mesothermal climates; its occurrence decreases with increasing latitude. *V. sempervirens* - a shade-tolerant/intolerant, submontane to montane, Western North American forb distributed more in the Pacific than the Cordilleran region. Occurs in maritime to submaritime cool mesothermal climates on moderately dry to fresh, nitrogen-medium soils; its occurrence decreases with increasing latitude and continentality.

Common in mossy understories on water-shedding sites; occasionally persist on cutover areas.

Viola palustris
Marsh violet

Violaceae
(Violet family)

A shade-intolerant, submontane to subalpine, circumpolar forb (transcontinental in North America). Occurs on wet to very wet soils within boreal, wet temperate, and cool mesothermal climates. Scattered in non-forested, semi-terrestrial communities and open-canopy forests on water-collecting sites (organic soils). Characteristic of wetlands.

Zigadenus venenosus
Meadow death-camas

Liliaceae
(Lily family)

A shade-intolerant, submontane to subalpine, Western North American herb distributed equally in the Pacific and Cordilleran regions. Occurs on very dry to moderately dry, nutrient-medium soils within temperate, cool semiarid, and mesothermal climates; its occurrence decreases with increasing elevation and precipitation and increases with temperature. Common in grassy communities on very shallow, often melanized soils on rock outcrops. Characteristic of moisture-deficient sites.

REFERENCES

Adamson, R.S. 1939. The classification of life-forms of plants. Bot. Rev. 5: 546-561.

Aichinger, E. 1967. Pflanzen als forstliche Standortsanzeigers. Österreichischer Agrarverlag, Wien.

Angove, K. and B. Bancroft. 1983. A guide to some common plants of the Southern Interior of British Columbia. Land Manage. Handbook No. 7, B.C. Min. For., Victoria, B.C.

Arbeitskreis Standortskartierung. 1978. Forstliche Standortsaufnahme. Arbeitskreis Standortskartierung in der Arbeitsgemeinschaft Forsteinrichtung. Landwirtschaftsverlag GmbH, Münster-Hilrup.

Archer, A.C. 1963. Some synecological problems in the alpine zone of Garibaldi Park. M.Sc. thesis. Dept. Biol. and Bot., Univ. British Columbia, Vancouver, B.C.

Arlidge, J.W.C. 1955. A preliminary classification and evaluation of Engelmann spruce - alpine fir forest at Bolean Lake, British Columbia. M.F. thesis, Fac. For., Univ. British Columbia, Vancouver, B.C.

Armson, K.A. 1977. Forest soils: properties and processes. Univ. Toronto Press, Toronto.

Baker, F.S. 1949. A revised tolerance table. J. For. 47: 179-181.

Bakuzis, E.V. 1969. Forestry viewed in an ecosystem perspective. pp. 189-258 in G.M. van Dyne (ed.) The ecosystem concept in natural resource management. Academic Press, New York.

Bakuzis, E.V. and V. Kurmis. 1978. Provisional list of synecological co-ordinates and selected ecographs of forest and other plant species in Minnesota. Dept. For. Resources, Coll. For. and Agric. Exp. Sta., Staff Series Paper No. 5, Univ. Minnesota, St. Paul, Minn.

Ballard, T.M. and R.E. Carter. 1986. Evaluating forest stand nutrient status. Land Manage. Rep. No. 20, B.C. Min. For., Victoria, B.C.

Beil, C.E. 1974. Forest associations of the southern Cariboo Zone, British Columbia. Syesis 7: 201-233.

Bell, M.A.M. 1971. Forest ecology. pp. 200-287 in Forestry Handbook for British Columbia. The Forest Club, Fac. For., Univ. British Columbia, Vancouver, B.C.

Black, C.A. 1968. Soil-plant relationships. John Wiley & Sons, Inc., New York.

Bradshaw, A.D., M.J. Chadwick, D. Jowett, and R.W. Snaydon. 1964. Experimental investigations into the mineral nutrition of several grass species. J. Ecol. 48: 665-676.

Brayshaw, T.C. 1970. The dry forests of southern British Columbia. Syesis 3: 17-43.

_____. 1976. Catkin bearing plants British Columbia. Occas. Pap. of British Columbia Prov. Museum No. 18, Prov. British Columbia, Victoria, B.C.

Bremner, J.M. 1965. Nitrogen availability indices. Agronomy 9: 1238-1255.

Brooke, R.C., E.B. Peterson, and V.J. Krajina. 1970. The subalpine Mountain Hemlock zone. Ecol. Western N. Amer. 2: 148-349.

Cain, S.A. 1950. Life-forms and phytoclimate. Bot. Rev. 16: 1-32.

Cajander, A.K. 1926. The theory of forest types. Acta For. Fenn., 2(3): 11-108.

Canada Depart. Agric. 1976. Glossary of terms in soil science. Publ. No. 1459, Research Branch, Can. Dept. Agric., Ottawa, Ont.

Canada Soil Survey Committee (CSSC). 1978. The Canadian system of soil classification. Can. Dept. Agric. Publ. No. 1646.

Carter, R.E. and K. Klinka. 1987. Relationships between Douglas-fir site index, actual evapotranspiration, and soil nutrients. A paper presented at the IUFRO Seminar: Management of Water and Water Relations to Increase Forest Growth. Canberra, Australia, 19-22 October.

Comeau, P.G., M.A. Comeau, and G.F. Utzig. 1982. A guide to plant indicators of moisture for southeastern British Columbia, with engineering interpretations. Land Manage. Handbook No. 5, B.C. Min. For., Victoria, B.C.

Cordes, L.D. 1972. An ecological study of the Sitka spruce forest on the west coast of Vancouver Island. Ph.D. thesis, Dept. Bot., Univ. British Columbia, Vancouver, B.C.

Coupé, R., C.A. Ray, A. Comeau, M.V. Ketcheson, and R.M. Annas. 1982. A guide to some common plants of the Skeena area, British Columbia. Land Manage. Handbook No. 4, B.C. Min. For., Victoria, B.C.

Courtin, P.J., K. Klinka, M.C. Feller and J.P. Demaerschalk. 1987. An approach to quantitative classification of nutrient regimes of forest soils. Can. J. Bot. (in print).

Curtis, J.T. 1959. The vegetation of Wisconsin - an ordination of plant communities. Univ. Wisconsin Press, Madison, Wisconsin.

Dahl, E. 1956. Rondane Mountain vegetation in South Norway and its relation to the environment. Norske Vidensk-Akad. Oslo, Mat-Naturv. Kl., Skr. 3: 1-374.

Daubenmire, R.F. 1968. Plant communities. Harper & Row, Inc., New York.

_____. 1976. The use of vegetation in assessing the productivity of forest lands. Bot. Rev. 42: 115-143.

de Coulon, J. 1923. *Nardus stricta.* Etude physilogique, anatomique, et embryologique. Mém. Soc. Vaud. Sci. Nat. 6: 245-332.

Egan, R.S. 1987. A fifth checklist of the lichen-forming lichenicolous and allied fungi of the continental United States and Canada. The Bryologist 90: 77-170.

Ellenberg, H. 1950. Umkrautgemeinschaften als Zeiger für Klima und Böden. Eugen Ulmer, Ludwigsburg.

———. 1952. Wiesen und Weiden und ihre standortliche Bewertung. Eugen Ulmer, Ludwigsburg.

———. 1953. Physiologisches und ökologisches Verhalten derselben Pflanzenarten. Ber. Deut. Botan. Ges. 65: 351-362.

———. 1974. Zeigerwerte der Gefässpflanzen Mitteleuropas. Scripta Geobotanica 9: 1-97.

———. 1982. Vegetation Mitteleuropas mit der Alpen in ökologischer Sicht. 3. Aufl., Verlag Eugen Ulmer, Stuttgart.

Emanuel, J. 1986. A vegetation classification program (F405:VTAB). Fac. For., Univ. British Columbia, Vancouver, B.C. (mimeographed).

Garman, E.H. 1973. Pocket guide to trees and shrubs in British Columbia. Min. Lands, For. and Water Resources, For. Serv., Victoria, B.C.

Gessner, F. 1932. Die Enstehung und Vernichtung von Pflanzengesselschaften und Vogelnistplätzen. B.B.C. (Erg.-Bd). 9: 113-128.

Giles, D.G. 1983. Soil water regimes on a forested watershed. M.Sc. thesis, Dept. Soil Sci., Univ. British Columbia, Vancouver, B.C.

Giles, D.G., T.A. Black, and D.L. Spittlehouse. 1985. Determination of growing season soil water deficits on a forested slope using water balance analysis. Can. J. For. Res. 15: 105-114.

Green, R.N., P.J. Courtin, K. Klinka, R.J. Slaco, and C.A. Ray. 1984. Site diagnosis, tree species selection, and slashburning guidelines for the Vancouver Forest Region, Land Manage. Handbook No. 8, B.C. Min. For., Victoria, B.C

Griffith, B.G. 1960. Growth of Douglas-fir at the University of British Columbia Research Forest. Forestry Bulletin No. 2, Fac. For., Univ. British Columbia, Vancouver, B.C.

Hale, M.E. 1969. The lichens. Wm. C. Brown Company Publishers. Dubuque, Iowa.

Hale, M.E. and W.L. Culberson. 1970. A fourth checklist of the lichens of the continental United States and Canada. The Bryologist 73: 499-543.

Haskin, L.L. 1977. Wild flowers of the Pacific coast. Dover Publications Inc., New York.

Hesselman, H. 1917. Studien über Nitratbildung im natürlichen Böden und ihre Bedeutung im pflanzensoziologischer Hinsicht. Medd. Skofsforsoksanst., Stockh. 13-14: 297-527.

Hitchcock, C.L. and A. Cronquist. 1973. Flora of the Pacific Northwest. Univ. Washington Press, Seattle, Washington.

Hitchcock, C.L., A. Cronquist, M. Ownbey, and J.W. Thompson. 1955-1969. Vascular plants of the Pacific Northwest. Univ. Washington Press, Seattle, Washington. 1(1969); 2(1964); 3(1961); 4(1959); 5(1955).

Hosie, R.C. 1979. Native trees of Canada. 7th ed., Can. For. Serv., Dept. Env., Supply and Serv. Canada, Ottawa, Ont.

Hultén, E. 1968. Flora of Alaska and neighboring territories. Stanford Univ. Press, Stanford, California.

Ireland, R.R., G.T. Brassand, W.B. Scholfield, and D.H. Vitt. 1987. Checklist of the mosses of Canada II. Lindbergia 13: 1-62.

Jeglum, J.K. 1971. Plant indicators of pH and water levels in peatlands at Candle Lake, Saskatchewan. Can. J. Bot. 49: 1661-1676.

Jenny, H. 1941. Factors of soil formation. McGraw-Hill Book Co., New York.

Kabzems, R.D. 1985. Quantitative classification of soil nutrient regimes of some mesothermal Douglas-fir ecosystems. M.Sc. thesis, Fac. For., Univ. British Columbia, Vancouver, B.C.

Kabzems, R.D. and K. Klinka. 1987. Initial quantitative classification of soil nutrient regimes. I. Soil properties, II. Relationships between soils, vegetation, and forest productivity. Can. J. For. Res. 17:1565-1571.

Kenkel, N.C. 1987. Trends and interrelationships in boreal wetland vegetation. Can. J. Bot. 65: 12-122.

Klinka, K. 1976. Ecosystem units - their classification, mapping and interpretation in the University of British Columbia Research Forest. Ph.D. thesis, Fac. For., Univ. British Columbia, Vancouver, B.C.

_____. 1977. Guide for tree species selection and prescribed burning in the Vancouver Forest District. B.C. Min. For., Vancouver, B.C.

Klinka, K., M.C. Feller, and L.E. Lowe. 1981. Characterization of the most productive ecosystems of growth of *Pseudotsuga menziesii* var. *menziesii* in southwestern British Columbia. Supplement to Land Manage. Rep. No. 6, B.C. Min. For. , Victoria, B.C.

Klinka, K., M.C. Feller, and R.K. Scagel. 1982. Characterization of the most productive ecosystems for the growth of Engelmann spruce (*Picea engelmannii* Parry ex Engelm.) in southwestern British Columbia. Land Manage. Rep. No. 9, B.C. Min. For., Victoria, B.C.

Klinka, K., R.N. Green, P.J. Courtin, and F.C. Nuszdorfer. 1984. Site diagnosis, tree species selection, and slashburning guidelines for the Vancouver Forest Region. Land Manage. Rep. No. 25, B.C., Min. For., Victoria, B.C.

Klinka, K., R.N. Green, R.L. Trowbridge, and L.E. Lowe. 1981. Taxonomic classification of humus forms in ecosystems of British Columbia. Land Manage. Rep. No. 8, B.C. Min. For., Victoria, B.C.

Klinka, K., F.C. Nuszdorfer, and L. Skoda. 1979. Biogeoclimatic units of central and southern Vancouver Island. Min. For., Victoria, B.C.

Kojima, S. and V.J. Krajina. 1975. Vegetation and environment of the Coastal Western Hemlock zone in Strathcona Provincial Park, British Columbia, Canada. Syesis 8(Suppl. 1): 1-123.

Krajina, V.J. 1965. Biogeoclimatic zones in British Columbia. Ecol. Western N. Amer. 1: 1-17.

_____. 1969. Ecology of forest trees in British Columbia. Ecol. Western N. Amer. 2(1): 1-146.

Krajina, V.J., K. Klinka, and E.B. Peterson. 1986. Ecology of vascular plants of British Columbia. Dept. Bot., Univ. British Columbia, Vancouver, B.C. (unpublished manuscript).

Krajina, V.J., K. Klinka, and J. Worrall. 1982. Distribution and ecological character-istics of trees and shrubs of British Columbia. Fac. For., Univ. British Columbia, Vancouver, B.C.

Kubiena, W. 1953. The soils of Europe. Thomas Murby and Co., London.

Kuramoto, R.T. 1965. Plant associations and succession in the vegetation of the sand dunes of Long Beach, Vancouver Island. M.Sc. thesis, Dept. Bot., Univ. British Columbia, Vancouver, B.C.

Landolt, E. 1977. Ökologische Zeigerwerte zur Schweizer Flora. Veröffentlichungen des Geobotanischen Institutes der Eidg. Techn. Hochshule, 64 Heft, Zürich.

Lawton, E. 1971. Moss flora of the Pacific Northwest. The Hattori Botanical Laboratory. Nichinan, Miyazaki, Japan.

Livingston, B.C. and F. Shreve. 1921. The distribution of vegetation in the United States, as related to climatic conditions. Carnegie Inst. Wash. Publ. 285: 1-585.

Long, J.N. and J. Turner. 1975. Above ground biomass of understory and overstory in an age sequence of four Douglas-fir stands. J. Appl. Ecol. 12: 179-188.

Lyons, C.P. 1952. Trees, shrubs and flowers to know in British Columbia. J.M. Dent & Sons (Canada) Ltd., Toronto, Ont.

Major, J. 1951. A functional factorial approach to plant ecology. Ecology 32: 392-412.

————. 1963. A climatic index to vascular plant activity. Ecology 44: 485-498.

————. 1969. Historical development of the ecosystem concept. pp. 9-22 in G.M. van Dyne (ed.) The ecosystem concept in natural resource management. Academic Press, New York.

————. 1977. California climate in relation to vegetation. pp. 11-74 in M.G. Barbour and J. Major (eds.) Terrestrial vegetation of California. John Wiley & Sons, New York.

McMinn, R.G. 1957. Water relations in the Douglas-fir region on Vancouver Island. Ph.D. thesis, Dept. Biol. and Bot., Univ. British Columbia, Vancouver, B.C.

————. 1961. Water relations and forest distribution in the Douglas-fir region on Vancouver Island. Canada Dept. Agric., Publ. No. 1901.

Meidinger, D.V. 1987. Recommended vernacular names for common plants of British Columbia. Research Report RR87002-HQ, B.C. Min. For., Research Branch, Victoria, B.C.

Meusel, H., E. Jäger, and E. Weinert. 1965. Vergleichende Chorologie der zentraleu-ropäischen Flora. Gustav Fischer Verlag, Jena.

Mezera, A. 1957. Flowering plants of our forests (in Czech). Statni Zemedelske Nakladatelstvi, Prague.

Minore, D. 1969. Yellow skunk-cabbage (*Lysichitum americanum* Hult & St. John) - an indicator of water table depth. Ecology 50(4): 737-739.

————. 1972. A classification of forest environments in the South Umpqua basin. Res. Pap. PNW-129, U.S. Dept. Agr., For. Serv., Portland, Oregon.

_____. 1979. Comparative autecological characteristics of northwestern tree species - a literature review. Gen. Tech. Rep. PNW-87, U.S. Dept. Agric., For. Serv., Portland, Oregon.

Minore, D. and R.E. Carkin. 1974. Proposed harvesting guides based upon an environmental classification in the South Umpqua Basin of Oregon. Res. Note PNW-232, U.S. Dept. Agric., For. Serv., Portland, Oregon.

_____. 1978. Vegetative indicators, soils, overstory canopy, and natural regeneration after partial cutting on the Dead Indian Plateau of southwestern Oregon. Res. Note PNW-316, U.S. Dept. Agric., For. Serv., Portland, Oregon.

Mueller-Dombois, D. 1959. The Douglas-fir forest associations on Vancouver Island in their initial stages of secondary succession. Ph. D. thesis, Dept. Biol. and Bot., Univ. British Columbia, Vancouver, B.C.

Mueller-Dombois, D. and H. Ellenberg. 1974. Aims and methods of vegetation ecology. John Wiley and Sons, Toronto.

Noble, W.J., T. Ahti, G.F. Otto, and I.M. Brodo. 1987. A second checklist and bibliography of the lichens and allied fungi of British Columbia. Syllogeus 61: 1-95.

Olsen, C. 1921. The ecology of *Urtica dioica*. J. Ecol. 9: 1-18.

Orloci, L. 1961. Forest types of the Coastal Western Hemlock zone. M.Sc. thesis, Dept. Biol. and Bot., Univ. British Columbia, Vancouver, B.C.

_____. 1964. Vegetational and environmental variations in the ecosystems of the Coastal Western Hemlock zone. Ph.D. thesis, Dept. Bot., Univ. British Columbia, Vancouver, B.C.

Persson, S. 1981. Ecological indicator values as an aid in the interpretation of ordination diagrams. J. Ecol. 69: 71-84.

Pigott, C.D. and K. Taylor. 1964. The distribution of some woodland herbs in relation to the supply of nitrogen and phosphorus in the soil. J. Ecol. 52 (Suppl.): 175-185.

Pogrebnyak, P.S. 1930. Über die Methodik der Standortsuntersuchungen in Verbindung mit den Waldtypen. pp. 455-471 *in* Proc. of the International Congress of For. Exp. Stations, Stockholm.

Pojar, J., R. Love, D. Meidinger, and R. Scagel. 1982. Some common plants of the Subboreal Spruce zone. Land Manage. Handbook No. 6, B.C. Min. For., Victoria, B.C.

Porsild, A.E. 1974. Rocky Mountains wild flowers. Natural History Series, No. 2, Botany Division, National Museum of Natural Sciences, National Museums of Canada, Ottawa, Ont.

Powers, R.F. 1980. Mineralizable soil nitrogen as an index of nitrogen availability to forest trees. Soil Sci. Soc. Am. J. 44: 1314-1320.

Pritchett, W.L. 1979. Properties and management of forest soils. John Wiley & Sons, New York.

Prusa, E. 1972. The application of development stages of forest communities. (in Czech). Lesnictvi 18(9): 793-831.

Prusa, E. and K. Pliva. 1969. Forest types as a framework for silviculture (in Czech). Statni Zemedelske Nakladatelstvi, Praha.

Raunkiaer, C. 1907. The life-forms of plants and their bearing on geography. pp. 2-104 in The Collected papers of C. Raunkiaer translated into English by H.G. Carter, A.G. Tansley, and M. Fausboll. 1934. Clarendon, Oxford.

Rees, W.J. and G.H. Sidrak. 1956. Plant nutrition on fly-ash. Plant and Soil 8: 141-159.

Revel, R.D. 1972. Phytogeocoenoses of the Sub-boreal Spruce biogeoclimatic zone in north central British Columbia. Ph.D. thesis, Dept. Bot., Univ. British Columbia, Vancouver, B.C.

Roemer, H. 1972. Forest vegetation and environments on the Saanich Peninsula. Ph.D. thesis, Dept. Biol., Univ. Victoria, Victoria, B.C.

Roy, R.J.J. 1984. Ordination and classification of immature forest ecosystems in the Cowichan Lake area, Vancouver Island. M.Sc. thesis, Fac. For., Univ. British Columbia, Vancouver, B.C.

Rowe, J.S. 1956. Uses of undergrowth plant species in forestry. Ecology 37(3): 461-473.

Schofield, W.B. 1969. Some common mosses of British Columbia. Handbook No. 28. B.C. Prov. Museum, Victoria, B.C.

Schuster, R.M. 1966-1974. The Hepaticae and Anthocerotae of North America. Columbia Univ. Press, New York. 1(1966); 2(1969); 3(1974).

Scoggan, H.J. 1978-1979. The flora of Canada. Publications in Botany No. 7 (1-4), National Museums of Canada, Ottawa. 1(1978); 2(1978); 3(1978); 4(1979).

Shimwell, D.W. 1972. The description and classification of vegetation. Univ. Washington Press, Seattle, Wash.

Shirley, H.L. 1943. Is tolerance the capacity to endure shade? J. For. 41: 339-345.

Schönhar, S. 1952. Untersuchungen über die Korrelation zwischen der floristichen Zusammensetzung der Bodenvegetation und der Bodenacidität sowie anderen chemischen Bodenfaktoren. Mitt. d. Ver. f. Forstl. Standortskartierung 2: 1-23.

_____. 1954. Die Bodenvegetation als Standortsweisers. Allg. Forst.-und Jagdzeitung 126: 259-266.

Shumway, J. and W.A. Atkinson. 1978. Predicting nitrogen fertilizer response in unthinned stands of Douglas-fir. Commun. Soil Sci. Plant Anal. 9: 529-539.

Sillinger, P. 1939. Biologie der nitrophilen Waldgesselschaften. Studia Botanica Cechica, Prague.

Smith, J.L., B.L. Neal, E.J. Owens, and G.O. Klock. 1981. Comparison of nitrogen mineralized under anaerobic and aerobic conditions for some agricultural and forest soils of Washington. Commun. Soil Sci. Plant Anal. 12: 997-1009.

Soil Survey Staff. 1975. Soil taxonomy. U.S. Dept. Agric., Soil Conservation Serv., Handbook No. 436, Washington, D.C.

Soper, J.H. and A.F. Szczawinski. 1976. Mount Revelstoke National Park wild flowers. Natural History Series No. 3, Botany Division, National Museum of Natural Sciences, National Museums of Canada, Ottawa, Ont.

Spittlehouse, D.L. and T.A. Black. 1981. A growing season water balance model applied to two Douglas-fir stands. Water Resour. Res. 17: 1651-1656.

Spurr, J.E. and B.V. Barnes. 1980. Forest ecology. John Wiley & Sons, New York.

Stotler, R. and B. Crandall-Stotler. 1977. A checklist of the liverworts and hornworts of North America. The Bryologist 80(3): 405-428.

Szczawinski, A.F. 1953. Corticolous and lignicolous plant communities in the forest associations of the Douglas-fir forest on Vancouver Island. Ph.D. thesis. Dept. Biol. and Bot., Univ. British Columbia, Vancouver, B.C.

_____. 1959. The orchids of British Columbia. Handbook No. 16, B.C. Prov. Museum, Victoria, B.C.

_____. 1962. The heather family of British Columbia. Handbook No. 19, B.C. Prov. Museum, Victoria, B.C.

Taylor, T.M.C. 1966. The lily family of British Columbia. Handbook No. 25, B.C. Prov. Museum, Victoria, B.C.

_____. 1971. The ferns and fern-allies of British Columbia. Handbook No. 12, B.C. Prov. Museum, Victoria, B.C.

_____. 1973. The rose family (*Rosaceae*) of British Columbia. Handbook No. 30, B.C. Prov. Museum, Victoria, B.C.

_____. 1974a. The pea family of British Columbia. Handbook No. 32, B.C. Prov. Museum, Victoria, B.C.

_____. 1974b. The figwort family of British Columbia. Handbook No. 33, B.C. Prov. Museum, Victoria, B.C.

Taylor, R.F. 1932. Plant indicators in S.E. Alaska. J. Forestry 30: 746.

Taylor, R.J. and G.W. Douglas. 1975. Mountain wild flowers of the Pacific Northwest. Binford & Mort, Portland, Oregon.

Taylor, R.L. and B. MacBryde. 1977. Vascular plants of British Columbia. Tech. Bull. No. 4, Univ. British Columbia Press, Vancouver, B.C.

Tisdale, S.L. and W.L. Nelson. 1975. Soil fertility and fertilizers. The MacMillan Company, New York.

Trewartha, G.T. 1968. An introduction to climate. 4th edition. McGraw-Hill, New York.

Tsiganov, D.N. 1983. Phyto-indication of ecological regimes in the coniferous-broadleaf zones. (In Russian). Izd. Nauka, Moscow.

Underhill, J.E. and C.C. Chuang. 1976. Wildflowers of Manning Park. B.C. Prov. Museum, Victoria, B.C.

Van Dyne, G.M. 1969. Preface. pp. vii-viii *in* G.M. Van Dyne (ed.) The ecosystem concept in natural resource management. Academic Press, New York.

Viktorov, S.V., E.A. Vostokova, and D.D. Vyshiukin. 1965. Some problems in the theory of geobotanical indicator research. pp. 1-4 *in* A.G. Chukishev (ed.) Plant indicators of soils, rocks, and subsurface waters. Consultants Bureau, New York.

Vitt, D.H., J.E. Marsh, and R.B. Bovey. 1988. A photographic field guide to the mosses, lichens and ferns of northwest North America. Lone Pine Publishing. Edmonton.

Wade, L.K. 1965. Vegetation and history of the *Sphagnum* bogs of the Tofino area, Vancouver Island. M.Sc. thesis, Dept. Bot., Univ. British Columbia, Vancouver, B.C.

Wali, M.K. and V.J. Krajina. 1973. Vegetation-environment relationships of some Subboreal Spruce zone ecosystems in British Columbia. Vegetatio 26: 237-381.

Walmsley, M., G. Utzig, T. Vold, D. Moon, and J. van Barnveld (eds.). 1980. Describing ecosystems in the field. B.C. Min. Env., RAB Tech. Pap. 2, B.C. Min. For. Land Manage. Rep. No. 7, Victoria, B.C.

Westhoff, V. and E. van der Maarel. 1978. The Braun-Blanquet approach. pp. 287-399 *in* R.H. Whittaker (ed.) Classification of plant communities. Dr. W. Junk b.v. Publishers, The Hague.

Whittaker, R.H. 1954. Plant populations and the basis of plant indication. Angewandte Pflanzensoziologie, Sonderfolge, Festschrift für Erwin Aichinger, Wien. 1: 183-206.

Whittaker, R.H. (ed.). 1978. Classification of plant communities. Dr. W. Junk b.v. Publishers, The Hague.

Wilde, S.A. 1954. Humus form: its genetic classification. Trans. Wis. Acad. Sci. 43: 127-163.

_____. 1966. A new systematic terminology of forest humus layers. Soil Sci. 101(5): 403-407.

_____. 1971. Forest humus: its classification on a genetic basis. Soil Sci. 3(1): 1-12.

Wilson, R.G. and W.R. Rouse. 1972. Moisture and temperature limits of the equilibrium evapotranspiration model. J. Appl. Meteorology 11(3): 436-442.

Zlatnik, A. 1970. Special forest botany. (in Czech). Statni Zemedelske Nakladatelstvi, Prague.

Zöttl, H. 1960. Die Mineralstickstoffanlieferung in Fichten-und Kiefernbeständen Bayerns. Forstwiss. Cbl. 79: 221-236.

APPENDIX I   Synopsis of indicator species groups and indicator values of species

Species	Life-form[1]	Indicator species group and indicator value[2]				Facultative indicator value[3]
		CLIM	MOIST	NITR	GSM	
Abies amabilis	CNTR	2 SBCM	4 FVM			
Abies grandis	CNTR	5 CTCM				
Abies lasiocarpa	CNTR	3 BCT				
Acer circinatum	DCSH	4 CM	4 FVM	3 R	2 MDML	
Acer glabrum	DCSH			3 R	2 MDML	
Acer macrophyllum	BLTR	4 CM	4 FVM	3 R	2 MDML	
Achillea lanulosa	FORB		2 VDMD	2 M	3 EMS	DIST
Achlys triphylla	FORB	4 CM		3 R	2 MDML	
Actaea rubra	FORB		4 FVM	3 R	2 MDML	
Adenocaulon bicolor	FORB	5 CTCM	3 MDF	3 R	2 MDML	
Adiantum pedatum	FERN	5 CTCM	4 FVM	3 R	2 MDML	CALC
Agropyron spicatum	GRAM	6 CTCSA	1 EDVD	3 R	2 MDML	ALKA
Agrostis aequivalvis	GRAM	4 CM	5 MW	2 M		
Aira caryophyllea	GRAM	4 CM	1 EDVD		4 VSS	
Aira praecox	GRAM	4 CM	1 EDVD	1 P	4 VSS	
Alectoria vancouverensis	LCHN	4 CM				
Allium acuminatum	FORB	5 CTCM	2 MD	2 M		
Allium cernuum	FORB			2 M		
Allotropa virgata	SAPR	5 CTCM	2 DMD	1 P	1 MOR	
Alnus rubra	BLTR	4 CM		3 R	2 MDML	DIST FGWT
Alnus sinuata	DCSH		4 FVM	3 R	2 MDML	DIST FGWT
Amelanchier alnifolia	DCSH		3 MDF	2 M		DIST
Anaphalis margaritacea	FORB				3 EMS	DIST
Andromeda polifolia	EGSH		6 WVW	1 P	5 SW	
Angelica genuflexa	FORB	5 CTCM	6 WVW	3 R	5 SW	
Antennaria neglecta	FORB	3 BCT	2 VDMD			
Apocynum androsaemifolium	FORB		2 VDMD	2 M	3 EMS	DIST
Aquilegia formosa	FORB		4 FVM	3 R	2 MDML	DIST
Aralia nudicaulis	FORB	3 BCT	4 FVM	3 R	2 MDML	
Arbutus menziesii	BLTR	4 CM	2 VDMD			
Arctostaphylos columbiana	EGSH	4 CM	2 VDMD	1 P		
Arctostaphylos uva-ursi	EGSH		2 VDMD	1 P	1 MOR	
Arnica cordifolia	FORB	3 BCT	3 MDF	2 M		
Arnica latifolia	FORB	1 ATB	4 FVM	2 M		SNOW
Aruncus dioicus	FORB		4 FVM	3 R	2 MDML	DIST
Asarum caudatum	FORB	5 CTCM	4 FVM	3 R	2 MDML	
Asplenium trichomanes	FERN				4 VSS	
Aster ciliolatus	FORB	3 BCT	3 MDF	2 M		
Aster conspicuus	FORB	3 BCT	3 MDF	3 R	2 MDML	ALKA
Athyrium filix-femina	FERN		5 MW	3 R	2 MDML	
Atrichum selwynii	MOSS	5 CTCM	4 FVM	3 R	3 EMS	DIST
Atrichum undulatum	MOSS	5 CTCM	4 FVM	3 R	3 EMS	DIST
Aulacomnium palustre	MOSS		6 WVW	2 M	5 SW	
Barbilophozia floerkei	LVRT	3 BCT	3 MDF	1 P	1 MOR	
Barbilophozia lycopodioides	LVRT	3 BCT	3 MDF	1 P	1 MOR	
Bartramia pomiformis	MOSS				4 VSS	
Bazzania tricrenata	LVRT	4 CM	4 FVM	1 P	1 MOR	

APPENDIX I   Synopsis of indicator species groups and indicator values of species (continued)

Species	Life-form[1]	Indicator species group and indicator value[2]				Facultative indicator value[3]
		CLIM	**MOIST**	**NITR**	**GSM**	
Blechnum spicant	FERN	2 SBCM	4 FVM	1 P	1 MOR	
Boschniakia hookeri	SAPR	4 CM		1 P	1 MOR	
Boykinia elata	FORB	4 CM	4 FVM	3 R	2 MDML	FLOOD
Brachythecium albicans	MOSS		3 MDF	2 M		
Bromus carinatus	GRAM		2 VDMD	2 M		
Bromus vulgaris	GRAM			3 R	2 MDML	
Calamagrostis canadensis	GRAM		5 MW	2 M		
Calamagrostis nutkaensis	GRAM	4 CM	5 MW	2 M		SALI
Calamagrostis rubescens	GRAM	3 BCT	2 VDMD	2 M		
Caltha biflora	FORB	2 SBCM	5 MW	3 R	2 MDML	
Caltha leptosepala	FORB	1 ATB	5 MW	3 R	2 MDML	SNOW
Calypogeia trichomanis	LVRT		4 FVM	1 P	1 MOR	
Calypso bulbosa	FORB		3 MDF	2 M		
Camassia leichtlinii	FORB	4 CM	3 MDF	3 R	2 MDML	
Camassia quamash	FORB	4 CM	3 MDF	3 R	2 MDML	
Campanula scouleri	FORB	4 CM	2 VDMD	1 P	1 MOR	
Cardamine breweri	FORB	5 CTCM	6 WVW	3 R	5 SW	
Cardamine nuttallii	FORB	4 CM	5 MW	3 R	2 MDML	
Carex anthoxanthea	GRAM	2 SBCM	6 WVW	2 M	5 SW	
Carex deweyana	GRAM		4 FVM	3 R	2 MDML	FGWT
Carex hendersonii	GRAM	4 CM	5 MW	3 R	2 MDML	FGWT
Carex inops	GRAM	4 CM	3 MDF	2 M		
Carex laeviculmis	GRAM		6 WVW	2 M	5 SW	
Carex livida	GRAM		6 WVW	2 M	5 SW	
Carex mertensii	GRAM		4 FVM	3 R	3 EMS	DIST
Carex obnupta	GRAM	4 CM	6 WVW	3 R	5 SW	
Carex rossii	GRAM		2 VDMD	2 M	4 VSS	
Carex sitchensis	GRAM	4 CM	6 WVW	3 R	5 SW	
Cassiope mertensiana	EGSH	1 ATB	3 MDF	1 P	1 MOR	SNOW
Cassiope stelleriana	EGSH	1 ATB	4 FVM	1 P	1 MOR	SNOW
Cassiope tetragona	EGSH	1 ATB	4 FVM	2 M		SNOW
Ceanothus sanguineus	DCSH		2 VDMD	2 M		DIST
Ceanothus velutinus	EGSH	6 CTCSA	3 MDF	2 M		DIST
Ceratodon purpureus	MOSS					DIST
Chamaecyparis nootkatensis	CNTR	2 SBCM				
Chimaphila menziesii	EGSH	4 CM	3 MDF	2 M		
Chimaphila umbellata	EGSH		2 VDMD	1 P	1 MOR	
Cinna latifolia	GRAM		4 FVM	3 R	2 MDML	FGWT
Circaea alpina	FORB		4 FVM	2 M		
Circaea pacifica	FORB	4 CM	4 FVM	3 R	2 MDML	FGWT
Cladina arbuscula	LCHN		1 EDVD	1 P	4 VSS	
Cladina impexa	LCHN		1 EDVD	1 P	4 VSS	
Cladina mitis	LCHN		1 EDVD	1 P	4 VSS	
Cladina rangiferina	LCHN		1 EDVD	1 P	4 VSS	
Cladina stellaris	LCHN	1 ATB	1 EDVD	1 P	4 VSS	

APPENDIX I   Synopsis of indicator species groups and indicator values of species (continued)

Species	Life-form[1]	Indicator species group and indicator value[2]				Facultative indicator value[3]
		CLIM	**MOIST**	**NITR**	**GSM**	
Cladonia bellidiflora	LCHN		2 VDMD	1 P	4 VSS	
Cladonia gracilis	LCHN		1 EDVD	1 P	4 VSS	
Cladothamnus pyroliflorus	DCSH	2 SBCM		1 P	1 MOR	
Claopodium crispifolium	MOSS	5 CTCM			4 VSS	CALC
Claytonia sibirica	FORB	5 CTCM	4 FVM	3 R	2 MDML	FGWT
Clintonia uniflora	FORB	3 BCT	3 MDF	1 P	1 MOR	
Collinsia parviflora	FORB		2 VDMD	2 M	4 VSS	
Conocephalum conicum	LVRT		5 MW	3 R		CALC
Coptis aspleniifolia	FORB	2 SBCM	4 FVM	1 P	1 MOR	
Coptis trifolia	FORB		5 MW	1 P	1 MOR	
Corallorhiza maculata	SAPR		3 MDF	1 P	1 MOR	
Corallorhiza mertensiana	SAPR		3 MDF	1 P	1 MOR	
Cornus canadensis	FORB	3 BCT		1 P	1 MOR	
Cornus nuttallii	BLTR	4 CM	3 MDF	3 R	2 MDML	
Cornus sericea	DCSH		5 VMW	3 R	2 MDML	FLOOD
Cornus unalaschkensis	FORB	2 SBCM	4 FVM	1 P	1 MOR	
Corylus cornuta	DCSH	5 CTCM	3 MDF	3 R	2 MDML	CALC
Crataegus douglasii	DCSH		5 VMW	3 R	2 MDML	
Cryptogramma crispa	FERN		2 VDMD	1 P	4 VSS	
Cystopteris fragilis	FERN		4 FVM	2 M		
Cytisus scoparius	DCSH	4 CM	2 VDMD	2 M	3 EMS	DIST
Danthonia intermedia	GRAM		2 VDMD	1 P	4 VSS	
Danthonia spicata	GRAM		2 VDMD	1 P	4 VSS	
Deschampsia caespitosa	GRAM		5 VMW	3 R		CALC DIST FGWT
Dicentra formosa	FORB	4 CM	4 FVM	3 R	2 MDML	DIST
Dicranum fuscescens	MOSS		2 VDMD	1 P	1 MOR	
Dicranum howellii	MOSS		3 MDF	1 P	1 MOR	
Dicranum pallidisetum	MOSS	1 ATB	3 MDF	1 P	1 MOR	
Dicranum tauricum	MOSS		2 VDMD	1 P	1 MOR	
Disporum hookeri	FORB		4 FVM	3 R	2 MDML	
Disporum smithii	FORB	4 CM	4 FVM	3 R	2 MDML	
Disporum trachycarpum	FORB	6 CTCS		3 R	2 MDML	
Dodecatheon hendersonii	FORB	4 CM	3 MDF	3 R	2 MDML	
Dodecatheon pulchellum	FORB		3 MDF	3 R	2 MDML	ALKA
Drosera rotundifolia	FORB		6 WVW	1 P	5 SW	
Dryopteris expansa	FERN		4 FVM	2 M		
Dryopteris filix-max	FERN		4 FVM	3 R	2 MDML	
Elymus glaucus	GRAM	5 CTCM	3 MDF	3 R	2 MDML	
Elymus hirsutus	GRAM		5 VMW	3 R	2 MDML	
Empetrum nigrum	EGSH			1 P	1 MOR	
Epilobium angustifolium	FORB			3 R		DIST
Epilobium latifolium	FORB	1 ATB		3 R		DIST
Equisetum arvense	FERN			2 M		DIST
Equisetum hyemale	FERN		4 FVM	3 R	3 EMS	CALC FLOOD
Equisetum sylvaticum	FERN	3 BCT	5 MW	1 P	1 MOR	
Equisetum telmateia	FERN	4 CM	4 FVM	3 R	3 EMS	FLOOD

APPENDIX I Synopsis of indicator species groups and indicator values of species (continued)

Species	Life-form[1]	Indicator species group and indicator value[2]				Facultative indicator value[3]
		CLIM	MOIST	NITR	GSM	
Erigeron peregrinus	FORB	1 ATB	4 FVM	3 R	2 MDML	SNOW
Eriophorum angustifolium	GRAM		6 WVW	1 P	5 SW	
Eriophyllum lanatum	FORB	5 CTCM	1 EDVD	2 M	4 VSS	
Erythronium oregonum	FORB	4 CM	3 MDF	2 M		
Erythronium revolutum	FORB	4 CM	4 FVM	3 R		FLOOD
Fauria crista-galli	FORB	2 SBCM	6 WVW	1 P	5 SW	
Festuca occidentalis	GRAM			1 P	1 MOR	
Festuca subulata	GRAM	4 CM	4 FVM	3 R	2 MDML	
Festuca subuliflora	GRAM	4 CM	4 FVM	3 R	2 MDML	
Fragaria vesca	FORB		3 MDF	2 M	3 EMS	DIST
Fragaria virginiana	FORB			2 M	3 EMS	CALC DIST
Fritillaria lanceolata	FORB	4 CM		3 R	2 MDML	
Galium aparine	FORB	5 CTCM		3 R	2 MDML	DIST
Galium triflorum	FORB		4 FVM	3 R	2 MDML	
Gaultheria humifusa	EGSH	1 ATB	4 FVM	1 P	1 MOR	SNOW
Gaultheria ovatifolia	EGSH	3 BCT	2 VDMD	1 P	1 MOR	
Gaultheria shallon	EGSH	4 CM		1 P	1 MOR	
Gentiana douglasiana	FORB	4 CM	6 WVW	1 P	5 SW	
Gentiana sceptrum	FORB	4 CM	6 WVW	2 M	5 SW	
Geocaulon lividum	FORB	3 BCT		1 P	1 MOR	
Geranium molle	FORB	4 CM	2 VDMD	2 M	4 VSS	DIST
Geum macrophyllum	FORB		4 FVM	3 R	3 EMS	DIST FGWT
Goodyera oblongifolia	FORB		3 MDF	1 P	1 MOR	
Gymnocarpium dryopteris	FERN		4 FVM	3 R	2 MDML	
Hemitomes congestum	PASA	5 CTCM	4 FVM	2 M		
Heracleum lanatum	FORB		4 FVM	3 R	2 MDML	FGWT
Herbertus aduncus	LVRT	4 CM				
Heuchera micrantha	FORB	4 CM		3 R	2 MDML	
Hieracium albiflorum	FORB		3 MDF		3 EMS	DIST
Hippuris montana	FORB	1 ATB	5 VMW	2 M		SNOW
Holcus lanatus	GRAM		4 FVM	2 M	3 EMS	DIST
Holodiscus discolor	DCSH	5 CTCM	2 VDMD	2 M		DIST
Homalothecium megaptilum	MOSS	4 CM	2 VDMD	1 P	1 MOR	
Hookeria acutifolia	MOSS	4 CM	5 VMW	1 P	1 MOR	
Hookeria lucens	MOSS	4 CM	5 VMW	1 P	1 MOR	
Huperzia selago	FERN		3 MDF	1 P	1 MOR	
Hylocomium splendens	MOSS			1 P	1 MOR	
Hypericum formosum	FORB		4 FVM	2 M		DIST
Hypochaeris radicata	FORB				3 EMS	DIST
Hypopithys lanuginosa	PASA		3 MDF	1 P	1 MOR	
Isopterygium elegans	MOSS		4 FVM	1 P		
Isothecium stoloniferum	MOSS	4 CM				
Juncus effusus	GRAM		5 VMW	2 M		DIST FGWT
Juncus ensifolius	GRAM		5 VMW	2 M		DIST FGWT
Juniperus scopulorum	EGSH	6 CTCSA	1 EDVD	2 M	4 VSS	ALKA
Juniperus sibirica	EGSH		2 VDMD	2 M		DIST

APPENDIX I Synopsis of indicator species groups and indicator values of species (continued)

Species	Life-form[1]	Indicator species group and indicator value[2]				Facultative indicator value[3]
		CLIM	MOIST	NITR	GSM	
Kalmia occidentalis	EGSH		6 WVW	1 P	5 SW	
Kindbergia oregana	MOSS	4 CM	3 MDF			
Kindbergia praelonga	MOSS		5 VMW	3 R	2 MDML	FGWT
Lathyrus nevadensis	FORB		3 MDF	3 R	2 MDML	
Lathyrus ochroleucus	FORB	3 BCT	3 MDF	3 R	2 MDML	
Ledum groenlandicum	EGSH		6 WVW	1 P	5 SW	
Lepidozia reptans	LVRT		4 FVM	1 P	1 MOR	
Leptarrhena pyrolifolia	FORB	1 ATB	5 VMW	2 M		
Letharia vulpina	LCHN	6 CTCSA				
Leucolepis menziesii	MOSS	4 CM	5 VMW	3 R	3 EMS	
Lilium columbianum	FORB		3 MDF	2 M		DIST
Linnaea borealis	FORB		3 MDF			
Listera caurina	FORB		4 FVM	2 M		
Listera convallarioides	FORB	5 CTCM	4 FVM	3 R	2 MDML	
Listera cordata	FORB			1 P	1 MOR	
Lobaria oregana	LCHN	4 CM				
Loiseleuria procumbens	EGSH		5 VMW	1 P	1 MOR	
Lomatium dissectum	FORB		2 VDMD	3 R	2 MDML	
Lonicera ciliosa	DCSH	5 CTCM	2 VDMD	2 M		
Lonicera hispidula	DCSH	4 CM	2 VDMD	2 M		
Lonicera involucrata	DCSH		5 VMW	3 R	2 MDML	FLOOD
Lonicera utahensis	DCSH	3 BCT		2 M		
Luetkea pectinata	FORB	1 ATB	4 FVM	2 M		SNOW
Lupinus arcticus	FORB	1 ATB		3 R		DIST
Lupinus nootkatensis	FORB			3 R		DIST
Luzula multiflora	GRAM	5 CTCM	1 EDVD	1 P	4 VSS	
Luzula parviflora	GRAM		4 FVM	2 M		
Lycopodium alpinum	FERN	1 ATB	3 MDF	2 M		
Lycopodium annotinum	FERN	3 BCT	3 MDF	2 M		
Lycopodium clavatum	FERN		3 MDF	1 P	1 MOR	
Lycopodium complanatum	FERN	3 BCT	3 MDF	1 P	1 MOR	
Lycopodium obscurum	FERN	3 BCT	4 FVM	1 P	1 MOR	
Lycopodium sitchense	FERN	1 ATB	4 FVM	1 P	1 MOR	SNOW
Lysichitum americanum	FORB		6 WVW	3 R	5 SW	
Madia madioides	FORB	4 CM	2 VDMD	2 M	3 EMS	DIST
Mahonia aquifolium	EGSH	6 CTCSA	2 VDMD	2 M		
Mahonia nervosa	EGSH	4 CM	3 MDF	2 M		
Maianthemum dilatatum	FORB	4 CM	5 VMW	3 R	2 MDML	FLOOD
Malus fusca	BLTR	4 CM	6 WVW	3 R	5 SW	
Marchantia polymorpha	LVRT		5 VMW		3 EMS	DIST
Melica subulata	GRAM	5 CTCM	4 FVM	3 R	2 MDML	
Menyanthes trifoliata	FORB		6 WVW	2 M	5 SW	
Menziesia ferruginea	DCSH		4 FVM	1 P		1 MOR
Mitella breweri	FORB	3 BCT	5 VMW	3 R	2 MDML	
Mitella nuda	FORB	3 BCT	4 FVM	2 M		
Mitella ovalis	FORB	4 CM	5 VMW	3 R	2 MDML	

APPENDIX I   Synopsis of indicator species groups and indicator values of species (continued)

Species	Life-form[1]	Indicator species group and indicator value[2]				Facultative indicator value[3]
		CLIM	**MOIST**	**NITR**	**GSM**	
Mitella pentandra	FORB		5 VMW	3 R	2 MDML	
Mnium spinulosum	MOSS	3 BCT	3 MDF	2 M		
Moehringia macrophylla	FORB	5 CTCM	3 MDF	3 R	2 MDML	
Moneses uniflora	FORB		4 FVM	2 M		
Monotropa uniflora	PASA		4 FVM	2 M		
Montia parvifolia	FORB	5 CTCM		2 M	4 VSS	
Mycelis muralis	FORB	4 CM	4 FVM	3 R	2 MDML	DIST
Myrica gale	DCSH		6 WVW	2 M	5 SW	
Nuphar polysepalum	FORB		6 WVW		5 SW	
Oemleria cerasiformis	DCSH	4 CM	4 FVW	3 R	2 MDML	FGWT
Oenanthe sarmentosa	FORB	4 CM	6 WVW	3 R	5 SW	
Oplopanax horridus	DCSH		5 VMW	3 R	2 MDML	
Orthilia secunda	FORB		3 MDF	1 P	1 MOR	
Osmorhiza chilensis	FORB		4 FVM	3 R	2 MDML	FGWT
Parnassia fimbriata	FORB	3 BCT	5 VMW	3 R	2 MDML	CALC
Paxistima myrsinites	EGSH	3 BCT	3 MDF	1 P	1 MOR	
Pedicularis bracteosa	FORB	3 BCT	4 FVM	2 M		
Pedicularis racemosa	FORB	3 BCT	3 MDF	2 M		
Pellia neesiana	LVRT		5 VMW	3 R	2 MDML	
Peltigera aphthosa	LCHN		2 VDMD	1 P	1 MOR	
Peltigera canina	LCHN		2 VDMD	2 M	1 MOR	
Peltigera membranacea	LCHN		2 VDMD	1 P	1 MOR	
Perideridia gairdneri	FORB	4 CM	3 MDF	2 M		
Petasites frigidus	FORB	1 ATB	5 VMW	3 R		DIST
Petasites palmatus	FORB		5 VMW	3 R		DIST FGWT
Phegopteris connectilis	FERN		4 FVM	3 R	2 MDML	CALC
Philadelphus lewisii	DCSH	6 CTCSA	3 MDF	2 M		
Philonotis fontana	MOSS		6 WVW	2 M	5 SW	
Phyllodoce empetriformis	EGSH	1 ATB	3 MDF	1 P	1 MOR	
Phyllodoce glanduliflora	EGSH	1 ATB	3 MDF	2 M		
Physocarpus capitatus	DCSH	5 CTCM	5 VMW	3 R	2 MDML	FLOOD
Picea engelmannii	CNTR	3 BCT				
Picea sitchensis	CNTR	4 CM		3 R		SALI
Pilophoron clavatus	LCHN	2 SBCM			4 VSS	
Pinus albicaulis	CNTR	1 ATB	3 MDF	2 M		
Pinus ponderosa	CNTR	6 CTCSA	2 VDMD	2 M		CALC
Plagiochila porelloides	LVRT		4 FVM	2 M		CALC
Plagiomnium insigne	MOSS	4 CM	5 VMW	3 R	3 EMS	
Plagiothecium undulatum	MOSS	4 CM	4 FVM	1 P	1 MOR	
Platanthera dilatata	FORB		6 WVW	2 M	5 SW	
Platanthera orbiculata	FORB	3 BCT	3 MDF	1 P	1 MOR	
Pleurozium schreberi	MOSS	3 BCT		1 P	1 MOR	
Pogonatum alpinum	MOSS	5 CTCM	4 FVM			
Pogonatum contortum	MOSS	4 CM	4 FVM	2 M	3 EMS	
Polypodium glycyrrhiza	FERN	4 CM			4 VSS	CALC
Polypodium scouleri	FERN	4 CM			4 VSS	

APPENDIX I   Synopsis of indicator species groups and indicator values of species (continued)

Species	Life-form[1]	Indicator species group and indicator value[2]				Facultative indicator value[3]
		CLIM	MOIST	NITR	GSM	
Polystichum braunii	FERN		4 FVM	3 R	2 MDML	
Polystichum lonchitis	FERN	1 ATB	3 MDF	2 M		CALC
Polystichum munitum	FERN	4 CM		3 R	2 MDML	
Polytrichum juniperinum	MOSS					DIST
Polytrichum piliferum	MOSS		1 EDVD	1 P	4 VSS	
Populus tremuloides	BLTR	3 BCT		3 R	2 MDML	
Populus trichocarpa	BLTR		4 FVM	3 R	2 MDML	FLOOD
Potentilla glandulosa	FORB		3 MDF	2 M		
Prenanthes alata	FORB	4 CM	4 FVM	3 R	2 MDML	FLOOD
Prunus virginiana	DCSH	6 CTCSA	3 MDF	3 R	2 MDML	ALKA
Pteridium aquilinum	FERN	5 CTCM				DIST
Pterospora andromeda	SAPR		2 VDMD	2 M		
Ptilium crista-castrensis	MOSS	3 BCT		1 P	1 MOR	
Pyrola asarifolia	FORB		3 MDF	2 M		
Pyrola chlorantha	FORB	3 BCT	3 MDF	2 M		
Pyrola picta	FORB	5 CTCM	3 MDF	2 M		
Quercus garryana	BLTR	4 CM	2 VDMD			
Ranunculus eschscholtzii	FORB	1 ATB	5 VMW	3 R	2 MDML	FLOOD
Ranunculus occidentalis	FORB			2 M		DIST
Ranunculus repens	FORB		5 VMW	3 R	3 EMS	DIST FGWT
Ranunculus uncinatus	FORB		4 FVM	3 R	3 EMS	DIST FGWT
Rhacomitrium canescens	MOSS		1 EDVD	1 P	4 VSS	
Rhacomitrium heterostichum	MOSS		1 EDVD	1 P	4 VSS	
Rhamnus purshianus	DCSH	5 CTCM	5 MW	3 R	2 MDML	FGWT
Rhizomnium glabrescens	MOSS	4 CM	4 FVM	2 M		
Rhizomnium magnifolium	MOSS		6 WVW	3 R	5 SW	
Rhizomnium nudum	MOSS	2 SBCM	5 MW	2 M		SNOW
Rhododendron albiflorum	DCSH	1 ATB	3 MDF	1 P	1 MOR	SNOW
Rhynchospora alba	GRAM		6 WVW	1 P	5 SW	
Rhytidiadelphus loreus	MOSS	2 SBCM	4 FVM	1 P	1 MOR	
Rhytidiadelphus triquetrus	MOSS			2 M		
Rhytidiopsis robusta	MOSS	3 BCT		1 P	1 MOR	
Ribes bracteosum	DCSH	4 CM	5 VMW	3 R	2 MDML	FLOOD
Ribes divaricatum	DCSH	4 CM	3 MDF	2 M		
Ribes lacustre	DCSH			3 R	2 MDML	
Ribes laxiflorum	DCSH		5 VMW	3 R	2 MDML	
Ribes lobbii	DCSH	4 CM	2 VDMD	2 M		
Ribes sanguineum	DCSH	4 CM	2 VDMD	2 M		
Rosa acicularis	DCSH	3 BCT	3 MDF	2 M		
Rosa gymnocarpa	DCSH		2 VDMD	2 M		
Rosa nutkana	DCSH		4 FVM	3 R	2 MDML	FGWT
Rubus idaeus	DCSH	3 BCT	4 FVM	3 R		DIST
Rubus laciniatus	EGSH	4 CM	4 FVM	3 R		DIST
Rubus leucodermis	DCSH		3 MDF	3 R		DIST
Rubus parviflorus	DCSH			3 R	2 MDML	
Rubus pedatus	FORB	3 BCT	4 FVM	1 P	1 MOR	

APPENDIX I   Synopsis of indicator species groups and indicator values of species (continued)

| Species | Life-form[1] | Indicator species group and indicator value[2] | | | | Facultative indicator value[3] |
		CLIM	MOIST	NITR	GSM	
Rubus pubescens	FORB	3 BCT	4 FVM	3 R	2 MDML	
Rubus spectabilis	DCSH	4 CM	5 VMW	3 R	2 MDML	FGWT
Rubus ursinus	DCSH	4 CM	3 MDF	2 M		DIST
Salix bebbiana	DCSH	3 BCT		2 M		DIST
Salix hookeriana	DCSH	4 CM	5 VMW	2 M		FLOOD DIST
Salix scouleriana	DCSH			2 M		DIST
Salix sitchensis	DCSH			2 M		DIST
Sambucus racemosa	DCSH		4 FVM	3 R	2 MDML	DIST
Sanguisorba canadensis	FORB		5 VMW			
Sanguisorba officinalis	FORB		5 VMW	2 M		DIST
Sanicula crassicaulis	FORB	4 CM	2 VDMD	3 R	2 MDML	
Sanicula graveolens	FORB		2 VDMD	3 R	2 MDML	
Satureja douglasii	FORB	4 CM	3 MDF	3 R	2 MDML	
Saxifraga ferruginea	FORB				4 VSS	
Saxifraga tolmiei	FORB	1 ATB	5 VMW	2 M	3 EMS	SNOW
Scapania bolanderi	LVRT	5 CTCM	4 FVM	1 P	1 MOR	
Scirpus microcarpus	GRAM	5 CTCM	6 WVW	3 R	5 SW	DIST
Sedum spathulifolium	FORB	4 CM	1 EDVD	1 P	4 VSS	
Selaginella wallacei	FERN		2 VDMD	1 P	4 VSS	
Senecio sylvaticus	FORB		3 MDF	3 R	3 EMS	DIST
Senecio triangularis	FORB		5 VMW	3 R	2 MDML	
Senecio vulgaris	FORB			3 R	3 EMS	DIST
Shepherdia canadensis	DCSH	3 BCT	2 VDMD	2 M		
Sibbaldia procumbens	EGSH	1 ATB	4 FVM	1 P	1 MOR	SNOW
Siphula ceratites	LCHN		6 WVW	1 P	1 MOR	
Sisyrinchium douglasii	FORB	4 CM	2 VDMD	2 M	4 VSS	
Smilacina racemosa	FORB	5 CTCM		3 R	2 MDML	
Smilacina stellata	FORB			3 R	2 MDML	
Sorbus scopulina	DCSH	3 BCT	3 MDF	2 M		
Sorbus sitchensis	DCSH	2 SBCM	3 MDF	1 P	1 MOR	
Sphagnum capillifolium	MOSS		6 WVW	1 P	5 SW	
Sphagnum fallax	MOSS		6 WVW	1 P	5 SW	
Sphagnum fuscum	MOSS		6 WVW	1 P	5 SW	
Sphagnum girgensohnii	MOSS		5 VMW	1 P	1 MOR	FGWT
Sphagnum papillosum	MOSS	4 CM	6 WVW	1 P	5 SW	
Sphagnum tenellum	MOSS	4 CM	6 WVW	1 P	5 SW	
Spiraea betulifolia	DCSH	3 BCT	2 VDMD	2 M		
Spiraea densiflora	DCSH	3 BCT	4 FVM	2 M		
Spiraea douglasii	DCSH	4 CM	5 VMW	2 M		DIST FGWT
Spiraea menziesii	DCSH		5 VMW	3 R		FGWT
Stachys cooleyae	FORB	4 CM	5 VMW	3 R	2 MDML	FLOOD
Stachys mexicana	FORB	4 CM	5 VMW	3 R	2 MDML	FLOOD
Stellaria calycantha	FORB	3 BCT	5 VMW			
Stellaria crispa	FORB		5 VMW	3 R	2 MDML	
Stenanthium occidentale	FORB	1 ATB	4 FVM	3 R	2 MDML	SNOW
Stereocaulon tomentosum	LCHN		1 EDVD	1 P	4 VSS	

APPENDIX I   Synopsis of indicator species groups and indicator values of species (concluded)

Species	Life-form[1]	Indicator species group and indicator value[2]				Facultative indicator value[3]
		CLIM	MOIST	NITR	GSM	
Streptopus amplexifolius	FORB		4 FVM	3 R	2 MDML	
Streptopus roseus	FORB		4 FVM	3 R	2 MDML	
Streptopus streptopoides	FORB	3 BCT	4 FVM	1 P	1 MOR	
Symphoricarpos albus	DCSH			3 R	2 MDML	FGWT
Symphoricarpos hesperius	DCSH	4 CM	2 VDMD	2 M		
Taxus brevifolia	CNTR	5 CTCM				
Tellima grandiflora	FORB	5 CTCM	4 FVM	3 R	2 MDML	FGWT
Thalictrum occidentale	FORB		4 FVM	3 R	2 MDML	FGWT
Thuja plicata	CNTR	5 CTCM				
Tiarella laciniata	FORB	4 CM	4 FVM	3 R	2 MDML	
Tiarella trifoliata	FORB		4 FVM	3 R	2 MDML	
Tiarella unifoliata	FORB	3 BCT	4 FVM	2 M		
Timmia austriaca	MOSS			2 M		CALC
Tofieldia glutinosa	FORB		6 WVW		5 SW	CALC
Tolmiea menziesii	FORB	4 CM	4 FVM	3 R	2 MDML	
Torreyochloa pauciflora	GRAM		6 WVW	3 R	5 SW	
Trautvetteria caroliniensis	FORB	5 CTCM	4 FVM	3 R	2 MDML	FGWT
Trichophorum cespitosum	GRAM	2 SBCM	6 WVW	1 P	5 SW	
Trientalis arctica	FORB		6 WVW	1 P	5 SW	
Trientalis latifolia	FORB	5 CTCM	3 MDF	2 M		
Trillium ovatum	FORB	4 CM	4 FVM	3 R	2 MDML	
Trisetum cernuum	GRAM		4 FVM	3 R	2 MDML	
Tsuga heterophylla	CNTR	5 CTCM			1 MOR[4]	
Tsuga mertensiana	CNTR	2 SBCM			1 MOR[4]	
Urtica lyallii	FORB		4 FVM	3 R		DIST
Usnea longissima	LCHN	4 CM				
Vaccinium alaskaense	DCSH	2 SBCM	4 FVM	1 P	1 MOR	
Vaccinium caespitosum	DCSH	3 BCT	4 FVM	1 P	1 MOR	
Vaccinium deliciosum	DCSH	2 SBCM	4 FVM	1 P	1 MOR	SNOW
Vaccinium membranaceum	DCSH	3 BCT	3 MDF	1 P	1 MOR	
Vaccinium myrtilloides	DCSH	3 BCT		1 P	1 MOR	
Vaccinium ovalifolium	DCSH		4 FVM	1 P	1 MOR	
Vaccinium ovatum	EGSH	4 CM	3 MDF	1 P	1 MOR	
Vaccinium oxycoccos	EGSH		6 WVW	1 P	5 SW	
Vaccinium parvifolium	DCSH	4 CM		1 P	1 MOR	
Vaccinium scoparium	DCSH	1 ATB		1 P	1 MOR	
Vaccinium uliginosum	DCSH		5 VMW	1 P	1 MOR	
Vahlodea atropurpurea	GRAM	1 ATB	4 FVM	2 M		SNOW
Valeriana scouleri	FORB		4 FVM	3 R	2 MDML	
Valeriana sitchensis	FORB		4 FVM	3 R	2 MDML	
Veratrum eschscholtzii	FORB		5 VMW	3 R	2 MDML	
Veronica americana	FORB		4 FVM	3 R	2 MDML	FLOOD
Viburnum edule	DCSH	3 BCT	4 FVM	3 R	2 MDML	FLOOD
Viburnum trilobum	DCSH	5 CTCM	4 FVM	3 R	2 MDML	
Vicia americana	FORB		3 MDF	2 M		
Viola adunca	FORB		2 VDMD	2 M		

APPENDIX I   Synopsis of indicator species groups and indicator values of species (concluded)

Species	Life-form[1]	Indicator species group and indicator value[2]				Facultative indicator value[3]
		CLIM	MOIST	NITR	GSM	
Viola glabella	FORB		5 VMW	3 R	2 MDML	FLOOD
Viola orbiculata	FORB		3 MDF	2 M		
Viola palustris	FORB		6 WVW		5 SW	
Viola sempervirens	FORB	4 CM	3 MDF	2 M		
Zigadenus venenosus	FORB		2 VDMD	2 M	4 VSS	

[1]   Symbols for life-forms given in Table 46 (page 64).

[2]   Symbols and numerical codes for ISGs and indicator values given in Table 4 (page 15) for climate; Table 12 (page 22) for soil moisture; Table 20 (page 30) for soil nitrogen; and Table 25 (page 37) for ground surface materials.

[3]   Symbols for facultative indicator values: ALKA - alkaline soils, CALC - calcium-rich soils, DIST - disturbed sites, FGWT - sites with prominently fluctuating groundwater table (the soil is wet or very wet in winter and slightly dry or fresh in summer), FLOOD - flooded sites (floodplains and stream-edge sites), SALI - saline soils, SNOW - late snow-melt sites.

[4]   Natural Regeneration

APPENDIX 2   Index of scientific and common names of species

Appendix 3 - Photo credits

page number, t - top, b - bottom

Boas, F. - 73t, 79b, 81b, 83t, 85t, 85b, 86t, 86b, 87t, 87b, 89t, 89b, 90b, 91b, 92t, 93t, 96b, 99b, 102t, 103b, 106t, 107t, 108t, 110t, 117t, 117b, 118b, 119b, 128b, 129b, 130t, 132b, 133b, 134t, 136t, 136b, 137t, 139b, 141b, 142t, 143t, 145t, 145b, 148t, 148b, 149t, 150b, 151b, 153t, 154t, 155t, 156b, 165b, 166t, 168b, 169t, 172b, 173t, 174b, 176b, 177t, 178t, 179b, 182t, 183b, 184b, 186t, 186b, 187t, 187b, 189b, 190t, 194b, 195t, 197b, 198t, 199t, 199b, 200t, 200b, 201t, 201b, 202t, 202b, 205b, 207b, 208b, 210t, 211b, 214b, 215t, 217b, 218b, 219b, 220t, 220b, 221b, 223b, 224b, 225b, 227b, 228b, 230t, 230b, 231t, 231b, 233t, 234t, 234b, 235t, 237t, 237b, 239t, 239b, 240b, 241t, 242b, 243b, 244t, 245t, 245b, 247t;

Ceska, A. - 68b, 69b, 70b, 71t, 72b, 74t, 76b, 78t, 80t, 80b, 81t, 83b, 92b, 93b, 94t, 94b, 95t, 95b, 96t, 97t, 97b, 98t, 98b, 99t, 101b, 105t, 105b, 108b, 109t, 109b, 110b, 111t, 112t, 112b, 111b, 115b, 116b, 120t, 121b, 122t, 131b, 134b, 137b, 138b, 139t, 140t, 140b, 141t, 142b, 146t, 152t, 157b, 158t, 159t, 160t, 161t, 161b, 162t, 162b, 164b, 165t, 167t, 172t, 173b, 175t, 177b, 178b, 180b, 184t, 190b, 203t, 203b, 205t, 206b, 208t, 209b, 211t, 212b, 213t, 214t, 215b, 216t, 219t, 221t, 226t, 229t, 232t, 238t, 240t, 241b, 246b, 247b, 248t;

Dickens, R. - 76t;

Hamilton, E. - 71b;

Kamloops Branch - 127t;

Kemp, A. - 118t;

Krajina, V.J. - 128t, 196b;

Long, R. - 70t, 74b, 88b, 119t, 156t, 163b, 170b, 179t, 188t, 233b;

Norton, R. - 69t, 90t, 100t, 100b, 125b, 138t, 174t, 207t, 209t, 210b, 213b, 232b, 236b, 243t;

Pojar, J. - 72t, 84t, 88t, 91t, 116t, 122b, 158b, 188b, 192b, 204t, 212t, 222t, 226b;

Ross, M. - 147b;

Turner, R. & N. - 150t, 206t, 227t;

Underhill, E.J. - 66t, 67t, 68t, 82t, 120b, 121t, 123b, 124t, 124b, 130b, 151t, 164t, 171t, 193t, 222b, 229b;

van Dieren, W. - 73b, 77t, 78b, 79t, 82b, 84b, 102b, 103t, 104b, 106b, 107b, 113b, 114t, 114b, 115t, 125t, 126t, 126b, 127b, 129t, 131t, 132t, 133t, 135t, 144t, 144b, 146b, 147t, 149b, 152b, 153b, 154b, 155b, 159b, 160b, 163t, 166b, 167b, 168t, 170t, 171b, 176t, 181b, 185t, 185b, 188t, 189t, 192t, 193b, 194t, 195b, 196t, 197t, 204b, 217t, 216b, 218t, 223t, 224t, 225t, 238b, 244b, 242t, 246t;

Wade, L.K. - 101t, 123t, 157t, 169b, 175b;

Walcott - 77b;

Appendix 3 - Photo credits (concluded)

Walker, R. - 143b;

Woolett, J. - 113t, 135b;

Worrall, J. - 66b, 67b, 75t, 75b, 104t, 181t, 182b, 183t, 191t, 191b, 198b, 228t, 235b, 236t.